Elements of Agricultural Chemistry

by Thomas Anderson

PREFACE.

The object of the present work is to offer to the farmer a concise outline of the general principles of Agricultural Chemistry. It has no pretensions to be considered a complete treatise on the subject. On the contrary, its aim is strictly elementary, and with this view I have endeavoured, as far as possible, to avoid unnecessary technicalities so as to make it intelligible to those who are unacquainted with the details of chemical science, although I have not hesitated to discuss such points as appeared essential to the proper understanding of any particular subject.

The rapid progress of agricultural chemistry, and the numerous researches prosecuted under the auspices of agricultural societies and private experimenters in this and other countries, render it by no means an easy task to make a proper selection from the mass of facts which is being daily accumulated. In doing this, however, I have been guided by a pretty intimate knowledge of the wants of the farmer, which has induced me to enlarge on those departments of the subject which bear more immediately on the every-day practice of agriculture; and for this reason the composition and properties of soils, the nature of manures, and the principles by which their application ought to be governed, have been somewhat minutely treated.

In all cases numerical details have been given as fully as is consistent with the limits of the work; and it may be right to state that a considerable number of the analyses contained in it have been made in my own laboratory, and that even when I have preferred to quote the results of other chemists, they have not unfrequently been confirmed by my own experiments.

UNIVERSITY OF GLASGOW, 1st November 1860.

CONTENTS.

INTRODUCTION 1

CHAPTER I.

THE ORGANIC CONSTITUENTS OF PLANTS.

Carbon ... Carbonic Acid ... Hydrogen ... Nitrogen ... Nitric Acid ... Ammonia ... Oxygen ... Sources whence obtained ... The Atmosphere ... The Soil ... Source of the Inorganic Constituents of Plants ... Manner in which the Constituents of Plants are absorbed 8

CHAPTER II.

THE PROXIMATE CONSTITUENTS OF PLANTS.

The Saccharine and Amylaceous Constituents ... Cellulose ... Incrusting Matter ... Starch ... Lichen Starch ... Inuline ... Gum ... Dextrine ... Sugar ... Mucilage ... Pectine and Pectic Acid ... Oily or Fatty Matters ... Margaric, Stearic, and Oleic Acids ... Wax ... Nitrogenous or Albuminous Constituents of Plants and Animals ... Albumen ... Fibrine ... Casein ... Diastase 40

CHAPTER III.

THE CHANGES WHICH TAKE PLACE IN THE FOOD OF PLANTS DURING THEIR GROWTH.

Changes occurring during Germination ... Changes during the After-Growth of the Plant ... Decomposition of Carbonic Acid ... Decomposition of Water ... Decomposition of Ammonia ... Decomposition of Nitric Acid 54

CHAPTER IV.

THE INORGANIC CONSTITUENTS OF PLANTS.

The Amount of Inorganic Matters in Different Plants ... The Relative Proportions of Ash in the Different Parts of Plants ... Influence of the Nature of the Soil on the Proportion of Mineral Matters in the Plant ... The Composition of the Ashes of Plants ... Classification of Different Plants 63

CHAPTER V.

THE SOIL--ITS CHEMICAL AND PHYSICAL CHARACTERS.

The Origin of Soils ... Composition of Crystalline and Sedimentary Rocks ... their Disintegration ... Chemical Composition of the Soil ... Fertile and Barren Soils ... Mechanical Texture of Soils ... Absorbent Action of Soils ... their Physical Characters ... Relation to Heat and Moisture ... The Subsoil ... Classification of Soils 83

CHAPTER VI.

THE IMPROVEMENT OF THE SOIL BY MECHANICAL PROCESSES.

Draining ... Its Advantageous Effects ... Subsoil and Deep Ploughing ... Improving the Soil by Paring and Burning ... Warping ... Mixing of Soils ... Chalking 137

CHAPTER VII.

THE GENERAL PRINCIPLES OF MANURING.

Fundamental Principles upon which Manures are applied ... Special and General Manures ... Importance of this distinction ... Views regarding the Theory of Manures ... Remarks on Special Manures ... Action of Manures on the Chemical and Physical Properties of a Soil ... Remarks on the Application of Manures 152

CHAPTER VIII.

THE COMPOSITION AND PROPERTIES OF FARM-YARD AND LIQUID MANURES.

Farm-yard Manure ... Urine ... Composition of ... Dung ... Composition of ... Farm-yard Manure ... Composition of ... Management of Dung-Heaps ... Box-feeding ... Fermentation and application of Manure ... Liquid Manure ... Composition and application of ... Sewage Manure ... Its composition and application 166

CHAPTER IX.

THE COMPOSITION AND PROPERTIES OF VEGETABLE MANURES.

Rape-Dust, Mustard, Cotton and Castor Cake ... Composition of various Oil-Cakes ... Malt-Dust, Bran, Chaff, etc. ... Straw and Saw-dust ... Manuring with Fresh Vegetable Matter ... Green Manuring ... Sea-Weed ... Composition of various Sea-Weeds ... Leaves ... Peat 195

CHAPTER X.

THE COMPOSITION AND PROPERTIES OF ANIMAL MANURES.

Guano, different varieties of ... Average composition of ... Division into Ammoniacal and Phosphatic ... Characters of ... Adulteration of ... Application of ... Pigeons' Dung ... Urate and Sulphated Urine ... Night-Soil and Poudrette ... Hair, Skin, Horn, Wool, etc. ... Blood ... Fish ... "Fish-Guano"--Bones 204

CHAPTER XI.

THE COMPOSITION AND PROPERTIES OF MINERAL MANURES.

Mineral Manures ... Sulphate and Muriate of Ammonia ... Sulphomuriate of Ammonia ... Ammoniacal Liquor ... Nitrates of Potash and Soda ... Muriate and Sulphate of Potash ... Chloride of Sodium, or Common Salt ... Carbonates of Potash and Soda ... Silicates of Potash and Soda ... Sulphate of Magnesia ... Phosphate of Lime ... Bone-ash ... Coprolites ... Apatite ... Sombrero Guano ... Superphosphates and Dissolved Bones ... Biphosphate of Lime or Soluble Phosphates ... Phospho-Peruvian Guano ... Lime ... Chalk ... Marl ... Application and Action of Lime on Soils ... Sulphate of Lime or Gypsum 226

CHAPTER XII.

THE VALUATION OF MANURES.

The Principle on which Manures are valued ... Its application to different simple and complex Manures ... Method of Calculation ... General Remarks 255

CHAPTER XIII.

THE ROTATION OF CROPS.

Its necessity explained ... Quantity of Mineral Matters in the produce of an Acre of Different Crops ... The Theory of Rotation 266

CHAPTER XIV.

THE FEEDING OF FARM STOCK.

The Principles of Feeding ... The Composition of different Animals in different stages of Fattening ... The Composition of the Food of Animals ... Milk ... The Principal Varieties of Cattle Food ... General Observations on Feeding 276

AGRICULTURAL CHEMISTRY.

INTRODUCTION.

That the phenomena of vegetation are dependent on certain chemical changes occurring in the plant, by which the various elements of its food are elaborated and converted into vegetable matter, was very early recognised by chemists; and long before the correct principles of that science were established, Van Helmont maintained that plants derived their nourishment from water, while Sir Kenelm Digby, Hook, Bradley, and others, attributed an equally exclusive influence to air, and enlarged on the practical importance of the conclusions to be deduced from their views. These opinions, which were little better than hypotheses, and founded on very imperfect chemical data, are mentioned by Jethro Tull, the father of modern agriculture, only to deny their accuracy; and he contended that the plants absorb and digest the finer particles of the earth, and attributed the success of the particular system of husbandry he advocated to the comminution of the soil, by which a larger number of its particles are rendered sufficiently small to permit their ready absorption by the roots. Popular opinion at that time was in favour of the mechanical rather than the chemical explanation of agricultural facts, and Tull's work had the effect of confirming this opinion, and turning attention away from the application of chemistry to agriculture. Indeed, no good results could have followed its study at that time, for chemistry, especially in those departments bearing more immediately on agriculture, was much too imperfect, and it was only towards the close of the last century, when Lavoisier established its true principles, that it became possible to pursue it with any prospect of success.

Very soon after Lavoisier's system was made known, Lord Dundonald published his "Treatise on the Intimate Connexion between Chemistry and Agriculture," in which the important bearings of the recent chemical discoveries on the practice of agriculture were brought prominently under the notice of the farmer, and almost at the same time De Saussure commenced those remarkable researches, which extended over a long series of years, and laid the foundation of almost all our accurate knowledge of the chemistry of vegetation. Saussure traced with singular care and accuracy the whole phenomena of the life of plants, and indicated the mode in which the facts he established might be taken advantage of in improving the cultivation of the soil. But neither his researches, nor Lord Dundonald's more direct appeal to the farmer, excited the attention they deserved, or produced any immediate effect on the progress of agriculture. It was not till the year 1812 that the interest of practical men was fairly awakened by a course of lectures given by Sir Humphrey Davy, at the instance of Sir John Sinclair, who was at that time president of the Board of Agriculture. In these lectures, written with all the clearness and precision which characterised their author's style, the results of De Saussure's experiments were for the first time presented to the farmer in a form in which they could be easily understood by him, the conclusions to which they led were distinctly indicated, and a number of useful practical suggestions made, many of which have been adopted into every-day practice, and become so thoroughly incorporated with it, that their scientific origin has been altogether forgotten. A lively interest was excited by the publication of Davy's work, but it soon died out, and the subject lay in almost complete abeyance for a considerable number of years. Nor could any other result be well expected, for at that time agriculture was not ripe for chemistry, nor chemistry ripe for agriculture. The necessities of a rapidly increasing population had not yet begun to compel the farmer to use every means adapted to increase the amount of production to its utmost limit; and though the fundamental principles of chemistry had been established, its details, especially in that department which treats of the constituents of plants and animals, were very imperfectly known. It is not surprising, therefore, that matters should have remained almost unchanged for the comparatively long period of nearly thirty years. Indeed, with the exception of the investigation of soils by Sch ü ler, and some other inquiries of minor importance, and which, in this country at least, excited no attention on the part of the agriculturist, nothing was done until the year 1840, when Liebig published his treatise on Chemistry, in its application to Agriculture and Physiology.

Saussure's researches formed the main groundwork of Liebig's treatise, as they had before done for Davy's; but the progress of science had supplied many new facts which confirmed the opinions of the older chemists in most respects, and enabled Liebig to generalise with greater confidence, and illustrate more fully the principles upon which chemistry ought to be applied to agriculture. Few works have ever produced a more profound impression. Written in a clear and forcible style, dealing with scientific truths in a bold and original manner, and producing a strong impression, as well by its earnestness as by the importance of its conclusions, it was received by the agricultural public with the full conviction that the application of its principles was to be immediately followed by the production of immensely increased crops, and by a rapid advance in every branch of practical agriculture. The disappointment of these extravagant expectations, which chemists themselves foresaw, and for which they vainly attempted to prepare the agriculturist, was followed by an equally rapid reaction; and those who had embraced Liebig's views, and lauded them as the commencement of a new era, but who had absurdly expected an instantaneous effect, changed their opinion, and contemned, as strongly as they had before supported, the application of chemistry to agriculture.

That this effect should have been produced is not unnatural; for practical men, having at that time little or no knowledge of chemistry, were necessarily unable to estimate its true position in relation to agriculture, and forgetting that this department of science was still in its early youth, and burthened with all the faults and errors of youth, they treated it as if it were already perfect in all its parts. Neither could they distinguish between the fully demonstrated scientific truths, and the uncertain, though probable conclusions deduced from them; and when the latter, as occasionally happened, proved to be at variance with practice, it is not surprising: that this should have produced a feeling of distrust on the part of persons incapable, from an imperfect, and still oftener from no knowledge of science, of drawing the line of demarcation, which Liebig frequently omitted to do, between the positive fact and the hypothetical inference, which, however probable, is, after all, merely a suggestion requiring to be substantiated by experiment. This omission, which the scientific reader can supply for himself, becomes a source of serious misapprehension in a work addressed to persons unacquainted with science, who adopt indiscriminately both the facts and the hypotheses of the author. And this is no doubt the cause of the vary different estimation in which the work of the Giessen Professor was held by scientific and practical men.

Liebig's treatise was followed, in the year 1844, by the publication of Boussingault's Economic Rurale, a work winch excited at the time infinitely less interest than Liebig's, although it is really quite as important a contribution to scientific agriculture. It is distinguished by entering more fully into the special details of the application of chemistry to agriculture, and contains the results of the author's numerous searches both in the laboratory and the field. Boussingault possesses the qualification, at present somewhat rare, of combining a thorough knowledge of practical agriculture with extended scientific attainments; and his investigations, which have been made with direct reference to practice, and their results tested in the field, are the largest and most valuable contribution to the exact data of scientific agriculture which has yet been made public.

The year 1844 was also distinguished by the foundation of the Agricultural Chemistry Association of Scotland, an event of no small importance in the history of scientific agriculture. That association was instituted through the exertions of a small number of practical farmers, for the purpose of pursuing investigations in agricultural chemistry, and affording to its members assistance in all matters connected with the cultivation of the soil, and has formed the model of similar establishments in London, Dublin, and Belfast, as well as in Germany; and it is peculiarly creditable to the intelligence and energy of the practical farmers of Scotland, that with them commenced a movement, which has already found imitators in so many quarters, and conferred such great benefits on agriculture. Within the last ten or twelve years, and mainly owing to the establishment of agricultural laboratories, great progress has been made in accumulating facts on which to found an accurate knowledge of the principles of agricultural chemistry, and the number of chemists who have devoted themselves to this subject has considerably increased, though still greatly less than its exigencies require.

Notwithstanding all that has recently been done, it must not be forgotten that we have scarcely advanced beyond the threshold, and that it is only by numerous and frequently repeated experiments that it is possible to arrive at satisfactory results. Agricultural inquiries are liable to peculiar fallacies due to the perturbing influence of climate, season, and many other causes, the individual effects of which can only be eliminated with difficulty, and much error has been introduced, by hastily generalising from single experiments, in place of awaiting the results of repeated trials. Hence it is that the progress of scientific agriculture must necessarily be slow and gradual, and is not likely to be marked by any great or startling discoveries. Now that the relations of science to practice are better understood, the extravagant expectations at one time entertained have been abandoned, and, as a necessary consequence, the interest in agricultural chemistry has again increased, and the conviction daily gains ground that no one who wishes to farm with success, can afford to be without some knowledge of the scientific principles of his art.

CHAPTER I.

THE ORGANIC CONSTITUENTS OF PLANTS.

When the water naturally existing in plants is expelled by exposure to the air or a gentle heat, the residual dry matter is found to be composed of a considerable number of different substances, which have been divided into two great classes, called the organic and the inorganic, or mineral constituents of plants. The former are readily combustible, and on the application of heat, catch fire, and are entirely consumed, leaving the inorganic matters in the form of a white residuum or ash. All plants contain both classes of substances; and though their relative proportions vary within very wide limits, the former always greatly exceed the latter, which in many cases form only a very minute proportion of the whole weight of the plant. Owing to the great preponderance of the organic or combustible matters, it was at one time believed that the inorganic substances formed no part of the true structure of plants, and consisted only of a small portion of the mineral matters of the soil, which had been absorbed along with their organic food; but this opinion, which probably was never universally entertained, is now entirely abandoned, and it is no longer doubted that both classes of substances are equally essential to their existence.

Although they form so large a proportion of the plant, its organic constituents are composed of no more than four elements, viz.:--

Carbon. Hydrogen. Nitrogen. Oxygen.

The inorganic constituents are much more numerous, not less than thirteen substances, which appear to be essential, having been observed. These are--

Potash. Soda. Lime. Magnesia. Peroxide of Iron. Silicic Acid. Phosphoric Acid. Sulphuric Acid. Chlorine.

And more rarely

Manganese. Iodine. Bromine. Fluorine.

Several other substances, among which may be mentioned alumina and copper, have also been enumerated; but there is every reason to believe that they are not essential, and the cases in which they have been found are quite exceptional.

It is to be especially noticed that none of these substances occur in plants in the free or uncombined state, but always in the form of compounds of greater or less complexity, and extremely varied both in their properties and composition.

It would be out of place, in a work like the present, to enter into complete details of the properties of the elements of which plants are composed, which belongs strictly to pure chemistry, but it is necessary to premise a few observations regarding the organic elements, and their more important compounds.

Carbon.--When a piece of wood is heated in a close vessel, it is charred, and converted into charcoal. This charcoal is the most familiar form of carbon, but it is not absolutely pure, as it necessarily contains the ash of the wood from which it was made. In its purest form it occurs in the diamond, which is believed to be produced by the decomposition of vegetable matters, and it is there crystallized and remarkably transparent; but when produced by artificial processes, carbon is always black, more or less porous, and soils the fingers. It is insoluble in water, burns readily, and is converted into carbonic acid. Carbon is the largest constituent of plants, and forms, in round numbers, about 50 per cent of their weight when dry.

Carbonic Acid.--This, the most important compound of carbon and oxygen, is best obtained by pouring a strong acid upon chalk or limestone, when it escapes with effervescence. It is a colourless gas, extinguishing flame, incapable of supporting respiration, much heavier than atmospheric air, and slightly soluble in water, which takes up its own volume of the gas. It is produced abundantly when vegetable matters are burnt, as also during respiration, fermentation, and many other processes. It is likewise formed daring the decay of animal and vegetable matters, and is consequently evolved from dung and compost heaps.

Hydrogen occurs in nature only in combination. Its principal compound is water, from which it is separated by the simultaneous action of an acid, such as sulphuric acid and a metal, in the form of a transparent gas, lighter than any other substance. It is very combustible, burns with a pale blue flame, and is converted into water. It is found in all plants, although in comparatively small quantity, for, when dry, they rarely contain more than four or five per cent. Its most important compound is water, of which it forms one-ninth, the other eight-ninths consisting of oxygen.

Nitrogen exists abundantly in the atmosphere, of which it forms nearly four-fifths, or, more exactly, 79 per cent. It is there mixed, but not combined with oxygen; and when the latter gas is removed, by introducing into a bottle of air some substance for which the former has an affinity, the nitrogen is left in a state of purity. It is a transparent gas, which is incombustible and extinguishes flame. It is a singularly inert substance, and is incapable of directly entering into union with any other element except oxygen, and with that it combines with the greatest difficulty, and only by the action of the electric spark--a peculiarity which has very important bearings on many points we shall afterwards have to discuss. Nitrogen is found in plants to the extent of from 1 to 4 per cent.

Nitric Acid.--This, the most important compound of nitrogen and oxygen, can be produced by sending a current of electric sparks through a mixture of its constituents, but in this way it can be obtained only in extremely small quantity. It is much more abundantly produced when organic matters are decomposed with free access of air, in which case the greater proportion of their nitrogen combines with the atmospheric oxygen. This process, which is known by the name of nitrification, is greatly promoted by the presence of lime or some other substance, with which the nitric acid may combine in proportion as it is formed. It takes place, to a great extent, in the soil in India and other hot climates; and our chief supplies of saltpetre, or nitrate of potash, are derived from the soil in these countries, where it has been formed in this manner. The same change occurs, though to a much smaller extent, in the soil in temperate climates.

Ammonia is a compound of nitrogen and hydrogen, but it cannot be formed by the direct union of these gases. It is a product of the decomposition of organic substances containing nitrogen, and is produced when they are distilled at a high temperature, or allowed to putrefy out of contact of the air. In its pure state it is a transparent and colourless gas, having a peculiar pungent smell, and highly soluble in water. It is an alkali resembling potash and soda, and, like these substances, unites with the acids and forms salts, of which the sulphate and muriate are the most familiar. In these salts it is fixed, and does not escape from them unless they be mixed with lime, or some other substance possessing a more powerful affinity for the acid with which it is united.

Oxygen is one of the most widely distributed of all the elements, and, owing to its powerful affinities, is the most important agent in almost all natural changes. It is found in the air, of which it forms 21 per cent, and in combination with hydrogen, and almost all the other chemical elements. In the pure state it possesses very remarkable properties. All substances burn in it with greater brilliancy than they do in atmospheric air, and its affinity for most of the elements is extremely powerful. When diluted with nitrogen, it supports the respiration of animals; but in the pure state it proves fatal after the lapse of an hour or two. It is found in plants, in quantities varying from 30 to 36 per cent.

It is worthy of observation, that of the four organic elements, carbon only is fixed, and the other three are gases; and likewise, when any two of them unite, their compound is either a gaseous or a volatile substance. The charring of organic substances, which is one of their most characteristic properties, and constantly made use of by chemists as a distinctive reaction, is due to this peculiarity; for when they are heated, a simpler arrangement of their particles takes place, the hydrogen, nitrogen, and oxygen unite among themselves, and carry off a small quantity of carbon, while the remainder is left behind in the form of charcoal, and is only consumed when access of the external air is permitted.

Now, in order that a plant may grow, its four organic constituents must be absorbed by it, and that this absorption may take place, it is essential that they be presented to it in suitable forms. A seed may be planted in pure carbon, and supplied with unlimited quantities of hydrogen, nitrogen, oxygen, and inorganic substances, and it will not germinate; and a plant, when placed in similar circumstances, shows no disposition to increase, but rapidly languishes and dies. The obvious inference from these facts is, that these substances cannot be absorbed when in the elementary state, but that it is only after they have entered into certain forms of combination that they acquire the property of being readily taken up, and assimilated by the organs of the plant.

It was at one time believed that many different compounds of these elements might be absorbed and elaborated, but later and more accurate experiments have reduced the number to four--namely, carbonic acid, water, ammonia, and nitric acid. The first supplies carbon, the second hydrogen, the two last nitrogen, while all of them, with the exception of ammonia, may supply the plant with oxygen as well as with that element of which it is the particular source.

There are only two sources from which these substances can be obtained by the plant, viz. the atmosphere and the soil, and it is necessary that we should here consider the mode in which they may be obtained from each.

The Atmosphere as a source of the Organic Constituents of Plants.-- Atmospheric air consists of a mixture of nitrogen and oxygen gases, watery vapour, carbonic acid, ammonia, and nitric acid. The two first are the largest constituents, and the others, though equally essential, are present in small, and some of them in extremely minute quantity. When deprived of moisture and its minor constituents, 100 volumes of air are found to contain 21 of oxygen and 79 of nitrogen. Although these gases are not chemically combined in the air, but only mechanically mixed, their proportion is exceedingly uniform, for analyses completely corresponding with these numbers have been made by Humboldt, Gay-Lussac, and Dumas at Paris, by Saussure at Geneva, and by Lewy at Copenhagen; and similar results have also been obtained from air collected by Gay-Lussac during his ascent in a balloon at the height of 21,430 feet, and by Humboldt on the mountain of Antisano in South America at a height of 16,640 feet. In short, under all circumstances, and in all places, the relation subsisting between the oxygen and nitrogen is constant; and though, no doubt, many local circumstances exist which may tend to modify their proportions, these are so slow and partial in their operations, and so counterbalanced by others acting in an opposite direction, as to retain a uniform proportion between the main constituents of the atmosphere, and to prevent the undue accumulation of one or other of them at any one point.

No such uniformity exists in the proportion of the minor constituents. The variation in the quantity of watery vapour is a familiar fact, the difference between a dry and moist atmosphere being known to the most careless observer, and the proportions of the other constituents are also liable to considerable variations.

Carbonic Acid.--The proportion of carbonic acid in the air has been investigated by Saussure. From his experiments, made at the village of Chambeisy, near Geneva, it appears that the quantity is not constant, but varies from 3·5 to 5·5 volumes in 10,000; the mean being 4·5. These variations are dependent on different circumstances. It was found that the carbonic acid was always more abundant during the night than during the day--the mean quantity in the former case being 4·2, in the latter 3·8. The largest quantity found during the night was 5·4, during the day 5·. Heavy and continued rain diminishes the quantity of carbonic acid, by dissolving and carrying it down into the soil. Saussure found that in the month of July 1827, during the time when nine millimetres of rain fell, the average quantity of carbonic acid amounted to 5·8 volumes in 10,000; while in September 1829, when 254 millimetres fell, it was only 3·7. A moist state of the soil, which is favourable to the absorption of carbonic acid, also diminishes the quantity contained in the air, while, on the other hand, continued frosts, by retaining the atmosphere and soil in a dry state, have an opposite effect. High winds increase the carbonic acid to a small extent. It was also found to be greater over the cultivated lands than over the lake of Geneva; at the tops of mountains than at the level of the sea; in towns than in the country. The differences observed in all these cases, though small, are quite distinct, and have been confirmed by subsequent experimenters.

Ammonia.--The presence of ammonia in the atmosphere appears to have been first observed by Saussure, who found that when the sulphate of alumina is exposed to the air, it is gradually converted into the double sulphate of alumina and ammonia. Liebig more recently showed that ammonia can always be detected in rain and snow water, and it could not be doubted that it had been absorbed from the atmosphere. Experiments have since been made by different observers with the view of determining the quantity of atmospheric ammonia, and their results are contained in the subjoined table, which gives the quantity found in a million parts of air.

Kemp 3?800

{ 12 feet above the surface 3?000 Pierre { 25 feet do. do. 0?000

Graeger 0?230

{ By day 0?980 Fresenius { By night 0?690

{ { Maximum 0?317 { In Paris { Minimum 0?177 { { Mean 0?237 Ville { { { Maximum 0?276 { Environs { Minimum 0?165 { of Paris { Mean 0?210

Of these results, the earlier ones of Kemp, Pierre, and Graeger are undoubtedly erroneous, as they were made without those precautions which subsequent experience has shown to be necessary. Even those of the other observers must be taken as giving only a very general idea of the quantity of ammonia in the air, for a proportion so minute as one fifty-millionth cannot be accurately determined even by the most delicate experiments. For this reason, more recent experimenters have endeavoured to arrive at conclusions bearing more immediately upon agricultural questions, by determining the quantity of ammonia brought down by the rain. The first observations on this subject were made by Barral in 1851, and they have been repeated during the years 1855 and 1856 by Mr. Way. In 1853, Boussingault also made numerous experiments on the quantity of ammonia in the rain falling at different places, as well as in dew and the moisture of fogs. He found in the imperial gallon--

Grs. Rain { Paris 0?100 { Liebfrauenberg 0?350

Dew, Liebfrauenberg { Maximum 0?340 { Minimum 0?714

{ Liebfrauenberg 0?790 Fog { Paris 9?000

It thus appears that in Paris the quantity of ammonia in rain-water is just six times as great as it is in the country, a result, no doubt, due to the ammonia evolved during the combustion of fuel, and to animal exhalations, and to the same cause, the large quantity contained in the moisture of fogs in Paris may also be attributed. Barral and Way have made determinations of the quantity of ammonia carried down by the rain in each month of the year, the former using for this purpose the water collected in the rain-gauges of the Paris Observatory, and representing, therefore, a town atmosphere; the latter, that from a large rain-gauge at Rothamsted, at a distance from any town. According to Barral the ammonia annually deposited on an acre of land amounts to 12?8 lbs., a quantity considerably exceeding that obtained by Way, whose experiments being made at a distance from towns, must be considered as representing more accurately the normal condition of the air. His results for the years 1855 and 1856 are given below, along with the quantities of nitric acid found at the same time.

Nitric Acid.--The presence of nitric acid in the air appears to have been first observed by Priestley at the end of the last century, but Liebig, in 1825, showed that it was always to be found after thunder-storms, although he failed to detect it at other times. In 1851 Barral proved that it is invariably present in rain-water, and stated the quantity annually carried down to an acre of land at no less than 41?9 lbs. But at the time his experiments were made, the methods of determining very minute quantities of nitric acid were exceedingly defective, and Way, by the adoption of an improved process, has shown that the quantity is very much smaller than Barral supposed, and really falls short of three pounds. His results for ammonia, as well as nitric acid, are given in the subjoined table.

	Ammonia in Grains.		Total Nitrogen in Grains.		Nitric Acid in Grains.	
	1855.	1856.	1855.	1856.	1855.	1856.
January	230	1564	1244	5,005	1084	4,526
February	944	544	2337	4,175	2169	3,579
March	1102	866	4513	2,108	3995	1,945
April	325	1063	1141	8,614	1024	7,369
May	1840	3024	4206	18,313	3939	15,863
June	3303	2046	5574	4,870	5447	4,540
July	2680	1191	9620	2,869	8615	2,670
August	3577	2125	4769	4,214	4870	4,021
September	732	1756	3313	5,972	2917	5,373
October	4480	2075	7592	3,921	7414	3,767
November	1007	1371	3021	2,591	2749	2,489
December	664	2035	2438	4,070	2180	3,352

-----+--------+--------+ |Total in pounds for}| | | | | | |the whole year }| 2?8 | ?80 | 7?1 | 9?3 | 6?3 | 8?1 | +--------------------+-------+--------+----------------+--------+--------+

No attempts have been made to determine the proportion of nitric acid in air, but its quantity is undoubtedly excessively minute, and materially smaller than that of ammonia. At least this conclusion seems to be a fair inference from Way's researches, as well as the recent experiments of Boussingault on the proportion of nitric acid contained in rain, dew, and fog, made in a manner exactly similar to those on the ammonia, already quoted. According to his experiments an imperial gallon contains--

Grs. Rain. {Paris 0?708 {Liebfrauenberg 0?140

Dew. {Maximum 0?785 {Minimum 0?030

Fog. {Paris 0?092 {Liebfrauenberg 0?718

Although it thus appears that Barral's results have been only partially confirmed, enough has been ascertained to show that the quantity of ammonia and nitric acid in the air is sufficient to produce a material influence in the growth of plants. The large amount of these substances contained in the dew is also particularly worthy of notice, and may serve to some extent to explain its remarkably invigorating effect on vegetation.

Carburetted Hydrogen.--Gay-Lussac, Humboldt, and Boussingault have shown, that when the whole of the moisture and carbonic acid have been removed from the air, it still contains a small quantity of carbon and hydrogen; and Saussure has rendered it probable that they exist in a state of combination as carburetted hydrogen gas. No definite proof of this position has, however, as yet been adduced, and the function of the compound is entirely unknown. It is possible that the presence of carbon and hydrogen may be due to a small quantity of organic matter; but, whatever be its source, its amount is certainly extremely small.

Sulphuretted Hydrogen and Phosphuretted Hydrogen.--The proportion of these substances is almost infinitesimal; but they are pretty general constituents of the atmosphere, and are apparently derived from the decomposition of animal and vegetable matters.

The preceding statements lead to the important conclusion, that the atmosphere is capable of affording an abundant supply of all the organic elements of plants, because it not only contains nitrogen and oxygen in the free state, but also in those forms of combination in which they are most readily absorbed, as well as a large quantity of carbonic acid, from which their carbon may be derived. At first sight it may indeed appear that the quantity of the latter compound, and still more that of ammonia, is so trifling as to be of little practical importance. But a very simple calculation serves to show that, though relatively small, they are absolutely large, for the carbonic acid contained in the whole atmosphere amounts in round numbers to

2,400,000,000,000 tons,

and the ammonia, assuming it not to exceed one part in fifty millions, must weigh

74,000,000 tons,

quantities amply sufficient to afford an abundant supply of these elements to the whole vegetation of our globe.

The Soil as a Source of the Organic Constituents of Plants.--When a portion of soil is subjected to heat, it is found that it, like the plant, consists of a combustible and an incombustible part; but while in the plant the incombustible part or ash is small, and the combustible large, these proportions are reversed in the soil, which consists chiefly of inorganic or mineral matters, mixed with a quantity of combustible or organic substances, rarely exceeding 8 or 10 per cent, and often falling considerably short of this quantity.

The organic matter exists in the form of a substance called humus, which must be considered here as a source of the organic constituents of plants, independently of the general composition of the soil, which will be afterwards discussed.

The term humus is generic, and applied by chemists to a rather numerous group of substances, very closely allied in their properties, several of which are generally present in all fertile soils. They have been submitted to examination by various chemists, but by none more accurately than by Mulder and Herman, to whom, indeed, we owe almost all the precise information we possess on the subject. The organic matters of the soil may be divided into three great classes; the first containing those substances which are soluble in water; the second, those extracted by means of caustic potash; and the third, those insoluble in all menstrua. When a soil is boiled with a solution of caustic potash, a deep brown fluid is obtained, from which acids precipitate a dark brown flocculent substance, consisting of a mixture of at least three different acids, to which the names of humic, ulmic, and geic acids have been applied. The fluid from which they have been precipitated contains two substances, crenic and apocrenic acids, while the soil still retains what has been called insoluble humus.

The acids above named do not differ greatly in chemical characters, but they have been subdivided into the humic, geic, and crenic groups, which present some differences in properties and composition. They are compounds of carbon, hydrogen, and oxygen, and are characterised by so powerful an affinity for ammonia that they are with difficulty obtained free from that substance, and generally exist in the soil in combination with it. They are all products of the decomposition of vegetable matters in the soil, and are formed during their decay by a succession of changes, which may be easily traced by observing the course of events when a piece of wood or any other vegetable substance is exposed for a length of time to air and moisture. It is then found gradually to disintegrate with the evolution of carbonic acid, acquiring first a brown and finally a black colour. At one particular stage of the process it is converted into one or other of two substances, called humin and ulmin, both insoluble in alkalies, and apparently identical with the insoluble humus of the soil; but when the decomposition is more advanced the products become soluble in alkalies, and then contain humic, ulmic, and geic acids, and finally, by a still further progress, crenic and apocrenic acids are formed as the result of an oxidation occurring at certain periods of the decay.

The roots and other vegetable debris remaining in the soil undergo a similar series of changes, and form the humus, which is found only in the surface soil, that is to say, in the portion which is now or has at some previous period been occupied by plants, and the quantity of humus contained in any soil is mainly dependent on the activity of vegetation on it. Numerous analyses of humus compounds extracted from the soil have been made, and have served to establish a number of minor differences in the composition even of those to which the same name has been applied, due manifestly to the fact that their production is the result of a gradual decomposition, which renders it impossible to extract from the soil one pure substance, but only a variable mixture of several, so similar to one another in properties, that their separation is very difficult, if not impossible. For this reason great discrepancies exist in the statements made regarding them by different observers, but this is a matter of comparatively small importance, as their exact composition has no very direct bearing on agricultural questions, and it will suffice to give the names and chemical formul?of those which have been analysed and described,--

Ulmic acid from long Frisian turf $C_{40} H_{18} O_{16}$ Humic acid from hard turf $C_{40} H_{15} O_{15}$ Humic acid from arable soil $C_{40} H_{16} O_{16}$ Humic acid from a pasture field $C_{40} H_{14} O_{14}$ Geic acid $C_{40} H_{15} O_{17}$ Apocrenic acid $C_{48} H_{12} O_{24}$ Crenic acid $C_{24} H_{12} O_{16}$

It is only necessary to observe further, that these formul?indicate a close connection with woody fibre, and the continuous diminution of the hydrogen and increase of oxygen shows that they must have been produced by a gradually advancing decay.

The earlier chemists and vegetable physiologists attributed to the humus of the soil a much more important function than it is now believed to possess.

It was formerly considered to be the exclusive, or at least the chief source of the organic constituents of plants, and by absorption through the roots to yield to them the greater part of their nutriment. But though this view has still some supporters, among whom Mulder is the most distinguished, it is now generally admitted that humus is not a direct source of the organic constituents of plants, and is not absorbed as such by their roots, although it is so indirectly, in as far as the decomposition which it is constantly undergoing in the soil yields carbonic acid, which can be absorbed. The older opinion is refuted by many well-ascertained facts. As regards the exclusive origin of the carbon of plants from humus, it is easy to see that this at least cannot be true, for humus, as already stated, is itself derived solely from the decomposition of vegetable and animal matters; and if the plants on the earth's surface were to be supported by it alone, the whole of their substance would have to return to the soil in the same form, in order to supply the generation which succeeds them. But this is very far from being the case, for the respiration of animals, the combustion of fuel, and many other processes, are annually converting a large quantity of these matters into carbonic acid; and if there were no other source of carbon but the humus of the soil, the amount of vegetable life would gradually diminish, and at length become entirely extinct. Schleiden, who has discussed this subject very fully, has made an approximative calculation of the total quantity of humus on the earth's surface, and of the carbon annually converted into carbonic acid by the respiration of man and animals, the combustion of wood for fuel, and other minor processes; and he draws the conclusion that, if there were no other source of carbon except humus, the quantity of that substance existing in the soil would only support vegetation for a period of sixty years.

The particular phenomena of vegetation also afford abundant evidence that humus cannot be the only source of carbon. Thus Boussingault has shown that on the average of years, the crops cultivated on an acre of land remove from it about one ton more organic matter than they receive in the manure applied to them, although there is no corresponding diminution in the quantity of humus contained in the soil. An instance which leads still more unequivocally to the same conclusion is given by Humboldt. He states that an acre of land, planted with bananas, yields annually about 152,000 pounds weight of fruit, containing about 32,000 pounds, or almost exactly 14 tons of carbon; and as this production goes on during a period of twenty years, there must be withdrawn in that time no less than 280 tons of carbon. But the soil on an acre of land weighs, in round numbers, 1000 tons, and supposing it to contain 4 per cent of humus, the total weight of carbon in it would amount to little more than 20 tons.

It is obvious from these and many other analogous facts that humus cannot be the only or even a considerable source of the carbon of plants, although it is still contended by some chemists that it may be absorbed to a small extent. But even this is at variance with many well-known facts. For if humus were absorbed, it might be expected that vegetation would be most luxuriant on soils containing abundance of that substance, especially if it existed in a soluble and readily absorbable form; but so far from this being the case, nothing is more certain than that peat, in which these conditions are fulfilled, is positively injurious to most plants. On the other hand, our daily experience affords innumerable examples of plants growing luxuriantly in soils and places where no humus exists. The sands of the sea-shore, and the most barren rocks, have their vegetation, and the red-hot ashes which are thrown out by active volcanoes are no sooner cool than a crop of plants springs up on them.

The conclusions to be drawn from these considerations have been further confirmed by the direct experiments of different observers. Boussingault sowed peas, weighing 15?0 grains, in a soil composed of a mixture of sand and clay, which had been heated red-hot, and consequently contained no humus, and after 99 days' growth, during which they had been watered with distilled water, he found the crop to weigh 68?2 grains, so that there had been a fourfold increase. Similar experiments have been made by Prince Salm Horstmar, on oats and rape sown in a soil deprived of organic matter by ignition, in which they grew readily, and arrived at complete maturity. One oat straw attained a height of three feet, and bore 78 grains; another bore 47; and a third 28--in all 153. These when dried at 212?weighed 46?02 grains, and the straw 45? grains. The most satisfactory experiments, however, are those of Weigman and Polstorf, these observers having found that it was possible to obtain a two-hundred-fold produce of barley in an entirely artificial soil, provided care was taken to give it the physical characters of a fertile soil. They prepared a mixture of six parts of sand, two of chalk, one of white bole, and one of wood charcoal; to which was added a small quantity of felspar, previously fused with marble and some soluble salts, so as to imitate as closely as possible the inorganic parts of a soil, and in it they planted twelve barley pickles. The plants grew luxuriantly, reaching a height of three feet, and each bearing nine ears, containing 22 pickles. The grain of the twelve plants weighed 2040 grains.

These experiments show that plants can grow and produce seed when the most scrupulous care is taken to deprive them of every trace of humus. But Saussure has gone further, and shown that even when present, humus is not absorbed. He allowed plants of the common bean and the Polygonum Persicaria to grow in solutions of humate of potash, and found a very trifling diminution in the quantity of humic acid present; but the value of his experiments is invalidated by his having omitted to ascertain whether the diminution of humic acid which he observed was really due to absorption by the plant. This omission has been supplied by Weigman and Polstorf. They grew plants of mint (Mentha undulata) and of Polygonum Persicaria in solutions of humate of potash, and placed beside the glass containing the plant, another perfectly similar, and containing only the solution of humate of potash. The solution, which contained in every 100 grains, 0?48 grains of solid matter, consisting of humate of potash, etc. was found to become gradually paler, and at the end of a month, during which time the plants had increased by 6-1/2 inches, the quantity of solid matter in 100 grains had diminished to 0?32. But the solution contained in the other glass, and in which no plant had grown, had diminished to 0?36, so that the absorption could not have amounted to more than 0?04 grains for every 100 grains of solution employed. This quantity is so small as to be within the limits of error of experiment, and we are consequently entitled to draw the conclusion that humus, even under the most favourable circumstances, is not absorbed by plants.

But though not directly capable of affording nutriment to plants, it must not, on that account, be supposed that humus is altogether devoid of importance, for it is constantly undergoing decomposition in the soil, and thus becomes a source of carbonic acid which can be absorbed, and, as we shall afterwards more particularly see, it exercises very important functions in bringing the other constituents of the soil into readily available forms of combination.

It has been already observed that carbon, hydrogen, nitrogen, and oxygen, cannot be absorbed by plants when uncombined, but only in the forms of water, carbonic acid, ammonia, and nitric acid. It is scarcely necessary to detail the grounds on which this conclusion has been arrived at in regard to carbon and hydrogen, for practically it is of little importance whether they can be absorbed or not, as the former is rarely, the latter never, found uncombined in nature. Neither can there be any doubt that water and carbonic acid are the only substances from which these elements can be obtained. Every-day experience convinces us that water is essential to vegetation; and Saussure, and other observers, have shown that plants will not grow if they are deprived of carbonic acid, and that they actually absorb that substance abundantly from the atmosphere. The evidence for the non-absorption of oxygen lies chiefly in the fact that plants obtain, in the form of water and carbonic acid, a larger quantity of that element than they require, and in place of absorbing, are constantly exhaling it. The form in which nitrogen may be absorbed has given rise to much difference of opinion. In the year 1779, Priestley commenced the examination of this subject, and drew from his experiments the conclusion, that plants absorb the nitrogen of the air. Saussure shortly afterwards examined the same subject, and having found, that when grown in a confined space of air, and watered with pure water, the nitrogen of the plants underwent no increase, he inferred that they derived their entire supplies of that element from ammonia, or the soluble nitrogenous constituents of the soil or manure. Boussingault has since re-examined this question, and by a most elaborate series of experiments, in which the utmost care was taken to avoid every source of fallacy, he was led to the conclusion, that when haricots, oats, lupins, and cresses were grown in calcined pumice-stone, mixed with the ash of plants, and supplied with air deprived of ammonia and nitric acid, their nitrogen underwent no increase. It has been objected to these experiments, that the plants being confined in a limited bulk of air, were placed in an unnatural condition, and Ville has recently repeated them with a current of air passing through the apparatus, and found a slight increase in the nitrogen, due, as he thinks, to direct absorption. It is much more probable, however, that it depends on small quantities of ammonia or nitric acid which had not been completely removed from the air by the means employed for that purpose, for nothing is more difficult than the complete abstraction of these substances, and as the gain of nitrogen was only 0? grains, while 60,000 gallons of air, and 13 of water, were employed in the experiment, which lasted for a considerable time, it is reasonable to suppose that a sufficient quantity may have remained to produce this trifling increase.

While these experiments show that plants maintain only a languid existence when grown in air deprived of ammonia and nitric acid, and hence, that the direct absorption of nitrogen, if it occur at all, must do so to a very small extent, the addition of a very minute quantity of the former substance immediately produces an active vegetation and rapid increase in size of the plants. Among the most striking proofs of this are the experiments of Wolff, made by growing barley and vetches in a soil calcined so as to destroy organic matters, and then mixed with small quantities of different compounds of ammonia. He found that when the produce from the calcined soil was represented by 100, that from the different ammoniacal salts was--

Barley. Vetches.

Muriate of Ammonia 257? 176? Carbonate of Ammonia 123? 173? Sulphate of Ammonia 203? 125?

These experiments not only prove that ammonia can be absorbed, but they also indirectly confirm the statement already made, that humus is not necessary; for in some instances the produce was higher than that obtained from the uncalcined soil with the same manures, although it contained four per cent of humus.

On such experiments Liebig rests his opinion that ammonia is the exclusive source of the nitrogen of plants, and although he has recently admitted that it may be replaced by nitric acid, it is obvious that he considers this a rare and exceptional occurrence. The evidence, however, for the absorption of nitric acid appears to rest on as good grounds as that of ammonia, for experience has shown that nitrate of soda acts powerfully as a manure, and its effect must be due to the nitric acid, and not to the soda, for the other compounds of that alkali have no such effect. Wolff has illustrated this point by a series of experiments on the sunflower, of which we shall quote one. He took two seeds of that plant, and sowed them on the 10th May, in a soil composed of calcined sand, mixed with a small quantity of the ash of plants, and added at intervals during the progress of the experiment, a quantity of nitrate of potash, amounting in all to 17?3 grains. The plants were watered with distilled water, containing carbonic acid in solution, and the pot in which they grew was protected from rain and dew by a glass cover. On the 19th August one of the plants had attained a height of above 28 inches, and had nine fine leaves and a flower-bud; the other was about 20 inches high, and had ten leaves. On the 22d August, one of the plants having been accidentally injured, the experiment was terminated. The plants, which contained 103?6 grains of dry matter, were then carefully analysed, and the quantity of nitrogen contained in the soil after the experiment and in the seed was determined.

Grains. Nitrogen in the dry plants 1?37 } " remaining in the soil 0?97 } 2?34

" in the nitrate of potash 2?70 } " in the seeds 0?29 } 2?99 ------ Difference 0?35

Hence, the nitrogen contained in the plants must, in this instance, have been obtained entirely from the nitrate of potash, for the quantity contained in it and in the seeds is exactly equal to that in the plants and the soil, the difference of 0?3 grains being so small that it may be safely attributed to the errors inseparable from such experiments. For the sake of comparison, an exactly similar experiment was made on two seeds grown without nitrate of potash, and in this instance, after an equally long period of growth, the largest plant had only attained a height of 7? inches, and had three small pale and imperfectly developed leaves. They contained only 0?33 grains of nitrogen, while the seeds contained 0?32--indicating that, under these circumstances, there was no increase in the quantity of that element.

But, independently of these experimental results, it may be inferred from general considerations, that nitric acid must be one of the sources from which plants derive their nitrogen. It has been already stated, that the humus contained in the soil consists of the remains of decayed plants, and there is every reason to suppose that the primeval soil contained no organic matters, and that the first generation of plants must have derived the whole of their nitrogen from, the atmosphere. If, therefore, it be assumed that ammonia is the only source of the nitrogen of plants, it would follow, that as that substance cannot be produced by the direct union of its elements, the quantity of ammonia in the air could only remain undiminished in the event of the whole of the nitrogen of decaying plants returning into that form. But this is certainly not the case, for every time a vegetable substance is burned, part of its nitrogen is liberated in the free state, and in certain conditions of putrefaction, nitric acid is produced. Now, if ammonia be the only form in which nitrogen is absorbed, there must be a gradual diminution of the quantity contained in the air; and further, there must either be some continuous source of supply by which its quantity is maintained, or there must be some other substance capable of affording nitrogen in a form fitted for the maintenance of plant life. As regards the first alternative, it must be stated that we know of no source other than the decomposition of plants from which ammonia can be derived, and we are therefore compelled to adopt the second alternative, and to admit that there must be some other source of nitrogen, and it cannot be doubted, from what has been already stated, that it is from nitric acid only that it can be obtained.

It must be admitted, then, that carbonic acid, ammonia, nitric acid, and water, are the great organic foods of plants. But while they have afforded to them an inexhaustible supply of the last, the quantity of the other three available for food are limited, and insufficient to sustain their life for a prolonged period. It has been shown by Chevandrier, that an acre of land under beech wood accumulates annually about 1650 lb. of carbon. Now, the column of air resting upon an acre of land contains only about 15,500 lb. of carbon, and the soil may be estimated to contain 1 per cent., or 22,400 lb. per acre, and the whole of this carbon would therefore be removed, both from the air and the soil, in the course of little more than 23 years. But it is a familiar fact, that plants continue to grow with undiminished luxuriance year after year in the same soil, and they do so because neither their carbon nor their nitrogen are permanently absorbed; they are there only for a period, and when the plant has finished its functions, and dies, they sooner or later return into their original state. Either the plant decays, in which case its carbon and nitrogen pass more or less rapidly into their original state, or it becomes the food of animals, and by the processes of respiration and secretion, the same change is indirectly effected. In this way a sort of balance is sustained; the carbon, which at one moment is absorbed by the plant, passes in the next into the tissues of the animal, only to be again expired in that state in which it is fitted to commence again its round of changes.

But while there is thus a continuous circulation of these constituents through both plants and animals, there are various changes which tend to liberate in the free state a certain quantity both of the carbon and nitrogen of plants, and these being thus removed from the sphere of organic life, there would be a gradual diminution in the amount of vegetation at the earth's surface, unless this loss were counterbalanced by some corresponding source of gain. In regard to carbonic acid the most important source is volcanic action, but the loss of nitrogen, which is far more important and considerable, is restored by the direct combination of its elements. The formation of nitric acid during thunder storms has been long familiar; but it would appear from the recent experiments of Cl 鯨 z, which, should they be confirmed by farther enquiry, will be of much importance, that this compound is also produced without electrical action when air is passed over certain porous substances, saturated with alkaline and earthy compounds. Fragments of calcined brick and pumice stone were saturated with solution of carbonate of potash, with carbonates of lime and magnesia and other mixtures, and a current of air freed from nitric acid and ammonia passed over them for a long period, at the end of which notable quantities of nitric acid were detected.

Source of the Inorganic Constituents of Plants.--The inorganic constituents of plants being all fixed substances, it is sufficiently obvious that they can only be obtained from the soil, which, as we shall afterwards see, contains all of them in greater or less abundance, and has always been admitted to be the only substance capable of supplying them. The older chemists and physiologists, however, attributed no importance to these substances, and from the small quantities in which they are found in plants, imagined that they were there merely accidental impurities absorbed from the soil along with the humus, which was at that time considered to be their organic food. This opinion, sufficiently disproved by the constant occurrence of the same substances in nearly the same proportions, in the ash of each individual plant, has been further refuted by the experiments of Prince Salm Horstmar, who has established their importance to vegetation, by experiments upon oats grown on artificial soils, in each of which one inorganic constituent was omitted. He found that, without silica, the grain vegetated, but remained small, pale in colour, and so weak as to be incapable of supporting itself; without lime, it died when it had produced its second leaf; without potash and soda, it grew only to the height of three inches; without magnesia, it was weak and incapable of supporting itself; without phosphoric acid, weak but upright; and without sulphuric acid, though normal in form, the plant was feeble, and produced no fruit.

Manner in which the Constituents of Plants are absorbed.--Having treated of the sources of the elements of plants, it is necessary to direct attention to the mode in which they enter their system.

Water.--The absorption of water by plants takes place in great abundance, and is connected with many of the most important phenomena of vegetation. It is principally absorbed by the roots, and passes into the tissues of the plant, where a part of it is decomposed, and goes to the formation of certain of its organic compounds; while by far the larger quantity, in place of remaining in it, is again exhaled by the leaves. The extent to which this takes place is very large. Hales found that a sunflower exhaled in twelve hours about 1 lb. 5 oz. of water, but this quantity was liable to considerable variation, being greater in dry, and less in wet weather, and much diminished during the night.

Saussure made similar experiments, and observed that the quantity of water exhaled by a sunflower amounted to about 220 lb. in four months. The exhalation of plants has recently been examined with great accuracy by Lawes. His experiments were made by planting single plants of wheat, barley, beans, peas, and clover, in large glass jars capable of holding about 42 lb. of soil, and covered with glass plates, furnished with a hole in the centre for the passage of the stem of the plant. Water was supplied to the soil at certain intervals, and the jars were carefully weighed. The result of the experiments, continued during a period of 172 days, is given in the following table, which shows the total quantity of water exhaled in grains:--

Wheat 113,527 Barley 120,025 Beans 112,231 Peas 109,082 Clover, cut 28th June 55,093

It further appears, that the exhalation is not uniform, but increases during the active growth of the plant, and diminishes again when that period is passed. These variations are shown by the subjoined tables, of which the first gives the total exhalation, and the second the average daily loss of water during certain periods.

TABLE I.--Showing the Number of Grains of Water given off by the Plants during stated divisional Periods of their Growth.

Description of Plant.	9 Days. From Mar. 19 to Mar. 28.	31 Days. From Mar. 28. to Apr. 28.	27 Days. From Apr. 28. to May 25.	34 Days. From May 25 to June 28.	30 Days. From June 28. to July 28.	14 Days. From July 28. to Aug. 11.	27 Days. From Aug. 11. to Sept. 7.
Wheat	129	1268	4,385	40,030	46,060	15,420	6235
Barley	129	1867	12,029	37,480	45,060	17,046	6414
Beans	88	1854	4,846	30,110	58,950	12,626	3657
Pease	101	1332	2,873	36,715	62,780	5,281	...
Clover	400	1645	2,948	50,100	...	...	...

TABLE II.--Showing the average daily Loss of Water (in Grains) by the Plants, within several stated divisional Periods of their Growth.

Description of Plant	9 Days. From Mar. 19 to Mar. 28	31 Days. From Mar. 28 to Apr. 28	27 Days. From Apr. 28 to May 25	34 Days. From May 25 to June 28	30 Days. From June 28 to July 28	14 Days. From July 28 to Aug. 11	27 Days. From Aug. 11 to Sept. 7
Wheat	14?	40?	162?	1177?	1535?	1101?	230?
Barley	14?	60?	445?	1102?	1502?	1217?	237?
Beans	9?	59?	179?	885?	1965?	901?	135?
Peas	11?	42?	106?	1079?	2092?	377?	...
Clover	44?	53?	109?	1473?	...	...	...

Similar experiments were made with the same plants in soils to which certain manures had been added, and with results generally similar. Calculating from these experiments, we are led to the apparently anomalous conclusion that the quantity of water exhaled by the plants growing on an acre of land greatly exceeds the annual fall of rain; although it is obvious that of all the rain which falls, only a small proportion can be absorbed by the plants growing on the soil, for a large quantity is carried off by the rivers, and never reaches their roots. It has been calculated, for instance, that the Thames carries off in this way at least one-third of the annual rain that falls in the district watered by it, and the Rhine nearly four-fifths. Of course this large exhalation must depend on the repeated absorption of the same quantity of water, which, after being exhaled, is again deposited on the soil in the form of dew, and passes repeatedly through the plant. This constant percolation of water is of immense importance to the plant, as it forms the channel through which some of its other constituents are carried to it.

Carbonic Acid.--While the larger part of the water which a plant requires is absorbed by its roots, the reverse is the case with carbonic acid. A certain proportion no doubt is carried up through the roots by the water, which always contains a quantity of that gas in solution, but by far the larger proportion is directly absorbed from the air by the leaves. A simple experiment of Boussingault's illustrates this absorption very strikingly. He took a large glass globe having three apertures, through one of which he introduced the branch of a vine, with twenty leaves on it. With one of the side apertures a tube was connected, by means of which the air could be drawn slowly through the globe, and into an apparatus in which its carbonic acid was accurately determined. He found, in this way, that while the air which entered the globe contained 0?004 of carbonic acid, that which escaped contained only 0?001, so that three-fourths of the carbonic acid had been absorbed.

Ammonia and Nitric Acid.--Little is known regarding the mode in which these substances enter the plant. It is usually supposed that they are entirely absorbed by the roots, and no doubt the greater proportion is taken up in this way, but it is very probable that they may also be absorbed by the leaves, at least the addition of ammonia to the air in which plants are grown, materially accelerates vegetation. It is probable, however, that the rain carries down the ammonia to the roots, and there is no doubt that that derived from the decomposition of the nitrogenous matters in the soil is so absorbed.

Inorganic Constituents.--The inorganic constituents of course are entirely absorbed by the roots; and it is as a solvent for them that the large quantity of water continually passing through the plants is so important. They exist in the soil in particular states of combination, in which they are scarcely soluble in water. But their solubility is increased by the presence of carbonic acid contained in the water, and which causes it to dissolve, to some extent, substances otherwise insoluble. It is in this way that lime, which occurs in the soil principally as the insoluble carbonate, is dissolved and absorbed. And phosphate of lime is also taken up by water containing carbonic acid, or even common salt in solution. The amount of solubility produced by these substances is extremely small; but it is sufficient for the purpose of supplying to the plant as much of its mineral constituents as are required, for the quantity of water which, as we have already seen, passes through a plant is very large when compared with the amount of inorganic matters absorbed. It has been shown by Lawes and Gilbert, that about 2000 grains of water pass through a plant for every grain of mineral matter fixed in it, so that there is no difficulty in understanding how the absorption takes place.

It is worthy of notice, however, that the absorption of the elements of plants takes place even though they may not be in solution in the soil, the roots apparently possessing the power of directly acting on and dissolving insoluble matters; but a distinction must be drawn between this and the view entertained by Jethro Tull, who supposed that they might be absorbed in the solid state, provided they were reduced to a state of sufficient comminution. It is now no longer doubted that, whatever action the roots may exert, the constituents of the plant must be in solution before they can pass into it--experiment having distinctly shown that the spongioles or apertures through which this absorption takes place are too minute to admit even the smallest solid particle.

CHAPTER II.

THE PROXIMATE CONSTITUENTS OF PLANTS.

The substances absorbed by the plant, which are of simple composition, and contain only two elements, are elaborated within it, and converted into the many complicated compounds of which its mass is composed. Some of these, as, for example, the colouring matters of madder and indigo, the narcotic principle of the poppy, &c., are confined to a single species, or small group of plants, while others are found in all plants, and form the main bulk of their tissues. The latter are the only substances which claim notice in a treatise like the present. They have been divided into three great classes, of widely different properties, composition, and functions.

1st. The Saccharine and Amylaceous Constituents.--These substances are compounds of carbon, hydrogen, and oxygen, and all possess a certain degree of similarity in composition, the quantities of hydrogen and oxygen they contain being always in the proportion required to form water, so that they may be considered as compounds of carbon and water; not that it can be asserted that they actually do contain water, as such, for of that there is no evidence, but only that its elements are present in the proportion to form it.

Cellulose.--This substance forms the fundamental part of all plants. It is the principal constituent of woody fibre, and is found in a state of purity in the fibre of cotton and flax, and in the pith of plants; but in wood it is generally contaminated with another substance, which has received the name incrusting matter, because it is deposited in and around the cells of which the plant is in part composed. Cellulose is insoluble in all menstrua, but, when boiled for a long time with sulphuric acid, is converted into a substance called dextrine. Cellulose consists of--

From pith of Elder-tree. Spongioles of roots.

Carbon 43·7 43·0 Hydrogen 6·4 6·8 Oxygen 50·9 50·2 ------- ------- 100·0 100·0

It is represented chemically by the formula, $C_{24}H_{21}O_{21}$, which shows it to be a compound of 24 atoms of carbon with 21 of hydrogen and 21 of oxygen.

Incrusting matter.--Large quantities of this substance enter into the composition of all plants. Of its chemical nature little is known, as it cannot be obtained separate from cellulose, but it is analogous to that substance in its composition, and probably contains hydrogen and oxygen in the proportion to form water.

Starch.--Starch is one of the most abundant constituents of plants, and is found in most seeds, as those of the cereals and the leguminous plants; in the tubers of the potatoe, the bulbs of tulips, &c. &c. It is obtained by placing a quantity of wheat flour in a bag, and kneading it under a gentle stream of water. When the water is allowed to stand, it deposits the starch as a fine white powder, which, when examined by the microscope, is found to be composed of minute grains, formed of concentric layers deposited on one another. These grains vary considerably in size and structure in different plants; but in the same plant they are generally so much alike as to admit of their recognition by a practised observer. They were formerly believed to be composed of an external coating of a substance insoluble in water, and containing in their interior a soluble kernel; but this opinion has been refuted, and distinct evidence been brought to show that the exterior and interior of the globules are identical in chemical properties. Starch is insoluble in cold water, but by boiling, it dissolves, forming a thick paste. By long continued boiling with water containing a small quantity of acid, it is completely dissolved and converted into dextrine, and eventually into sugar. The same change is produced by the action of fermenting substances, such as the extract of malt; when heated in the dry state to a temperature of about 390 Fahr., it becomes soluble in cold water. It is distinguished by giving a brilliant blue compound with iodine. Starch contains--

Carbon 44?7 Hydrogen 6?8 Oxygen 49?5 ------ 100?0

and its composition is represented by the formula $C_{12}H_{10}O_{10}$, so that it differs but little from cellulose in composition, although its chemical functions in the plant are extremely different. It is connected with some of the most important changes which occur in the growing plants, and by a series of remarkable transformations is converted into sugar and other important compounds.

Lichen Starch is found in most species of lichens, and is distinguished from common starch by producing a green colour with iodine. Its composition is the same as that of ordinary starch.

Inuline.--The species of starch to which this name is given is characterised by its dissolving in boiling water, and giving a white pulverulent deposit in cooling. It is found in the tuber of the dahlia, in the dandelion, and some other plants. Its composition is identical with that of cellulose, and its formula is $C_{24}H_{21}O_{21}$.

Gum is excreted from various plants as a thick fluid, which dries up into transparent masses. Its composition is identical with that of starch. It dissolves readily in cold water, and is converted into sugar by long continued boiling with acids. Its properties are best marked in gum arabic, which is obtained from various species of acacia; that from other plants differs to some extent, although its chemical composition is the same.

Dextrine.--When starch is exposed to a heat of about 400? or when treated with sulphuric acid, or with a substance extracted from malt called diastase, it is converted into dextrine. It may also be obtained from cellulose by a similar treatment. The dextrine so obtained has the same composition as the starch from which it is produced, but its properties more nearly resemble those of gum. It plays a very important part in the process of germination, and may be converted into sugar on the one hand, and apparently also into starch on the other.

Sugar.--Under this name are included four or five distinct substances, of which the most important are, cane sugar, grape sugar, and the uncrystallisable sugar found in many plants.

Cane Sugar.--This variety of sugar, as its name implies, is found most abundantly in the sugar cane, but it occurs also in the maple, beet-root, and various species of palms, from all of which it is extracted on the large scale. It is extremely soluble in water, and can be obtained in large transparent prismatic crystals, as in common sugar-candy. It swells up, and is converted into a brown substance called caramel, when heated, and by contact with fermenting substances, yields alcohol and carbonic acid. It contains--

Carbon 42·2 Hydrogen 6·0 Oxygen 51·8 ------ 100·0

and its chemical formula is $C_{12}H_{11}O_{11}$.

Grape Sugar is met with in the grape, and most other fruits, as well as in honey. It is produced artificially when starch is boiled for a long time with sulphuric acid, or treated with a large quantity of diastase. It is less soluble in water than cane sugar, and crystallises in small round grains. Its composition, when dried at 284° is--

Carbon 40·0 Hydrogen 6·6 Oxygen 53·4 ------ 100·0

and its formula is $C_{12}H_{12}O_{12}$; but when crystallised it contains two equivalents of water, and is then represented by the formula $C_{12}H_{12}O_{12} + 2H_2O$.

The uncrystallisable sugar of plants is closely allied to grape sugar, and, so far as at present known, has the same composition, although, from the difficulty of obtaining it quite free from crystallised sugar, this is still uncertain.

Mucilage is the name applied to the substance existing in linseed, and in many other seeds, and which communicates to them the property of swelling up and becoming gelatinous when treated with water. It is found in a state of considerable purity in gum tragacanth and some other gums. Its composition is not known with absolute certainty, but it is either $C_{24}H_{19}O_{19}$, or $C_{12}H_{10}O_{10}$; and in the latter case it must be identical with starch and gum.

It will be observed that all the substances belonging to this class are very closely related in chemical composition, some of them, as starch and gum, though easily distinguished by their properties, being identical in constitution, while others only differ in the quantity of water, or of its elements which they contain. In fact, they may all be considered as compounds of carbon and water, and their relations are, perhaps, more distinctly seen when their formulæ are written so as to show this, as is done in the following table, in the second column of which those containing twelve equivalents of carbon are doubled, so as to make them comparable with cellulose:--

			Water.
Grape sugar,	$C_{12}H_{12}O_{12}$	$C_{24}H_{24}O_{24}$	C_{24} + 24
Cane sugar,	$C_{12}H_{11}O_{11}$	$C_{24}H_{22}O_{22}$	C_{24} + 22
Cellulose,	$C_{24}H_{21}O_{21}$	$C_{24}H_{21}O_{21}$	C_{24} + 21
Inuline,	$C_{24}H_{21}O_{21}$	$C_{24}H_{21}O_{21}$	C_{24} + 21
Starch,	$C_{12}H_{10}O_{10}$	$C_{24}H_{20}O_{20}$	C_{24} + 20
Dextrine,	$C_{12}H_{10}O_{10}$	$C_{24}H_{20}O_{20}$	C_{24} + 20
Gum,	$C_{12}H_{10}O_{10}$	$C_{24}H_{20}O_{20}$	C_{24} + 20
Mucilage,	$C_{12}H_{10}O_{10}$	$C_{24}H_{20}O_{20}$	C_{24} + 20

The relation between these substances being so close, it is not difficult to understand how one may be converted into another by the addition or subtraction of water. Thus, cellulose has only to absorb an equivalent of water to become grape sugar, or to lose an equivalent in order to be converted into starch, and we shall afterwards see that such changes do actually occur in the plant during the process of germination.

Pectine and Pectic Acid.--These substances are met with in many fruits and roots, as, for instance, in the apple, the carrot, and the turnip. They differ from the starch group in containing more oxygen than is required to form water along with their hydrogen; but their exact composition is still uncertain, and they undergo numerous changes during the ripening of the fruit.

2d. Oily or Fatty Matters.--The oily constituents of plants form a rather extensive group of substances all closely allied, but distinguished by minor differences in properties and constitution. Some of them are very widely distributed throughout the vegetable kingdom, but others are almost peculiar to individual plants. They are all compounds of carbon, hydrogen, and oxygen, and are at once distinguished from the preceding class, by containing much less oxygen than is required to form water with their hydrogen. The principal constituents of the fatty matters and oils of plants are three substances, called stearine, margarine, and oleine, the two former solids, the latter a fluid; and they rarely, if ever, occur alone, but are mixed together in variable proportions, and the fluidity of the oils is due principally to the quantity of the last which they contain. If olive oil be exposed to cold, it is seen to become partially solid; and if it be then pressed, a fluid flows out, and a crystalline substance remains; the former is oleine, though not absolutely pure, and the latter margarine. The perfect separation of these substances involves a variety of troublesome chemical processes; and when it has been effected, it is found that each of them is a compound of a peculiar acid, with another substance having a sweet taste, and which has received the name of glycerine, or the sweet principle of oil. Glycerine, as it exists in the fats, appears to be a compound of C_3H_2O, and its properties are the same from whatever source it is obtained. The acids separated from it are known by the names of margaric, stearic, and oleic acids.

Margaric Acid is best obtained pure by boiling olive oil with an alkali until it is saponified, and decomposing the soap with an acid, expressing the margaric acid, which separates, and crystallising it from alcohol. It is a white crystalline fusible solid, insoluble in water, but soluble in alcohol and in solutions of the alkalies. Its composition is--

Carbon 75?6 Hydrogen 12?9 Oxygen 11?5 ------ 100?0

and its formula $C_{34}H_{34}O_4$.

Stearic Acid.--Although this acid exists in many plants, it is most conveniently extracted from lard. It is a crystalline solid less fusible than margaric acid, but closely resembling it in its other properties. Its formula is $C_{36}H_{36}O_4$.

Oleic Acid.--Under this name two different substances appear to be included. It has been applied generally to the fluid acids of all oils, while it would appear that the drying and non-drying oils actually contain substances of different composition. The acid extracted from olive oil appears to have the formula $C_{36}H_{34}O_4$, while that from linseed oil is $C_{46}H_{38}O_6$, but this is still doubtful.

Other fatty acids have been detected in palm oil, cocoa-nut oil, &c. &c., which so closely resemble margaric and stearic acids as to be easily confounded with them. Though presenting many points of interest, it is unnecessary to describe them in detail here.

Wax is a substance closely allied to the oils. It consists of two substances, cerine and myricine, which are separated from one another by boiling alcohol, in which the former is more soluble. They are extremely complex in composition, the former consisting principally of an acid similar to the fatty acids, called cerotic acid, and containing $C_{54}H_{54}O_{4}$. The latter has the formula $C_{92}H_{92}O_{4}$. The wax found in the leaves of the lilac and other plants appears to consist of myricine, while that extracted from the sugar-cane is said to be different, and to have the formula $C_{48}H_{50}O_{2}$. It is probable that other plants contain different sorts of wax, but their investigation is still so incomplete, that nothing definite can be said regarding them. Wax and fats appear to be produced in the plant from starch and sugar; at least it is unquestionable that the bee is capable of producing the former from sugar, and we shall afterwards see that a similar change is most probably produced in the plant. The fatty matters contained in animals are identical with those of plants.

3d. Nitrogenous or Albuminous Constituents of Plants and Animals.--The nitrogenous constituents of plants and animals are so closely allied, both in properties and composition, that they may be most advantageously considered together.

Albumen.--Vegetable albumen is found dissolved in the juices of most plants, and is abundant in that of the potato, the turnip, and wheat. In these juices it exists in a soluble state, but when its solution is heated to about 150? it coagulates into a flocky insoluble substance. It is also thrown down by acids and alcohol. Coagulated albumen is soluble in alkalies and in nitric acid. Animal albumen exists in the white of eggs, the serum of blood, and the juice of flesh; and from all these sources is scarcely distinguishable in its properties from vegetable albumen.

It is a substance of very complicated composition, and chemists are not agreed as to the formula by which its constitution is to be expressed, a difficulty which occurs also with most of the other nitrogenous compounds.

The results of the analyses of albumen from different sources are however quite identical, as may be seen from those subjoined--

	From Wheat.	From Potatoes.	From Blood.	From White of Egg.
Carbon	53?	53?	53?	53?
Hydrogen	7?	7?	7?	7?
Nitrogen	15?	...	15?	15?
Oxygen	} {	...	22?	22?
Sulphur	} 23?	{ 0?7	1?	1?
Phosphorus	} {	...	0?	0?
	-----	-----	-----	
	100?	100?	100?	

Closely allied to vegetable albumen is the substance known by the name of glutin, which is obtained by boiling the gluten of wheat with alcohol. It appears to be a sort of coagulated albumen, with which its composition completely agrees.

Vegetable Fibrine.--If a quantity of wheat flour be tied up in a piece of cloth, and kneaded for some time under water, the starch it contains is gradually washed out, and there remains a quantity of a glutinous substance called gluten. When this is boiled with alcohol, the glutin above referred to is extracted, and vegetable fibrine is left. It dissolves in dilute potash, and on the addition of acetic acid is deposited in a pure state. Treated with hydrochloric acid, diluted with ten times its weight of water, it swells up into a jelly-like mass. When boiled or preserved for a long time under water, it cannot be distinguished from coagulated albumen.

Animal Fibrine exists in the blood and the muscles, and agrees in all its characters and composition with vegetable fibrine, as is shown by the subjoined analyses--

	Wheat Flour.	Blood.	Flesh.
Carbon	53?	52?	53?
Hydrogen	7?	6?	7?
Nitrogen	15?	15?	15?
Oxygen	23?	24?	23?
Sulphur	1?	1?	1?
	-----	-----	-----
	100?	100?	100?

Caseine.--Vegetable caseine exists abundantly in most plants, especially in the seeds, and remains in the juice after albumen has been precipitated by heat, from which it may be separated in flocks by the addition of an acid. It has been obtained for chemical examination, principally from peas and beans, and from the almond and oats. When prepared from the pea it has been called legumine, from almonds emulsine, and from oats avenine; but they are all three identical in their properties, although formerly believed to be different, and distinguished by these names. Vegetable caseine is best obtained by treating peas or beans with hot water, and straining the fluid. On standing, the starch held in suspension is deposited, and the caseine is retained in solution in the alkaline fluid; by the addition of an acid it is precipitated as a thick curd. Caseine is insoluble in water, but dissolves readily in alkalies; its solution is not coagulated by heat, but, on evaporation, becomes covered with a thin pellicle, which is renewed as often as it is removed.

Animal Caseine is the principal constituent of milk, and is obtained by the cautious addition of an acid to skimmed milk, by which it is precipitated as a thick white curd. It is also obtained by the use of rennet, and the process of curding milk is simply the coagulation of its caseine. It is soluble in alkalies, and precipitated from its solution by acids, and in all other respects agrees with vegetable caseine.

The composition of animal caseine has been well ascertained, but considerable doubt still exists as to that of vegetable caseine, owing to the difficulty of obtaining it absolutely pure. The analyses of different chemists give rather discordant results, but we have given those which appear most trustworthy--

From Peas. Carbon 50? 50? Hydrogen 6? 6? Nitrogen 16? 15? Oxygen 25? 23? Sulphur 0? 0? Phosphorus ... 2? ----- ----- 100? 100?

Other results differ considerably from these, and some observers have even obtained as much as eighteen per cent of nitrogen and fifty-three of carbon.

The composition of animal caseine differs from this principally in the amount of carbon. Its composition is--

Carbon 53? Hydrogen 7? Nitrogen 15? Oxygen 22? Sulphur 1? ----- 100?

The most cursory examination of these analytical numbers is sufficient to show that a very close relation subsists between the different substances just described. Indeed, with the exception of vegetable caseine, they may be said all to present the same composition; and, as already mentioned, there are analyses of it which would class it completely with the others. While, however, the quantities of carbon, hydrogen, nitrogen, and oxygen are the same, differences exist in the sulphur and phosphorus they contain, and which, though very small in quantity, are indubitably essential to them. Much importance has been attributed to these constituents by various chemists, and especially by Mulder, who has endeavoured to make out that all the albuminous substances are compounds of a substance to which he has given the name of proteine, with different quantities of sulphur and phosphorus. The composition of proteine, according to his newest experiments, is--

Carbon 54? Hydrogen 7? Nitrogen 16? Oxygen 21? Sulphur 1? ----- 100?

and is exactly the same from whatever albuminous compound it is obtained. Although the importance of proteine is probably not so great as Mulder supposed, it affords an important illustration of the close similarity of the different substances from which it is obtained, the more especially as there is every reason to believe that the different albuminous compounds are capable of changing into one another, just as starch and sugar are mutually convertible; and the possibility of this change throws much light on many of the phenomena of nutrition in plants and animals. Indeed, it would seem probable that these compounds are formed from their elements by plants only, and are merely assimilated by animals to produce the nitrogenous constituents they contain.

Diastase is the name applied to a substance existing in malt, and obtained by macerating that substance with cold water, and adding a quantity of alcohol to the fluid, when the diastase is immediately precipitated in white flocks. It is produced during the malting process, and is not found in the unmalted barley. Its chemical composition is unknown, but it is nitrogenous, and is believed to be produced by the decomposition of gluten. If a very small quantity of diastase be mixed with starch suspended in hot water, the starch is found gradually to dissolve, and to pass first into the state of dextrine, then into that of sugar. The change thus effected takes place also in a precisely similar manner in the plant, diastase being produced during the process of germination of all seeds and tubers, for the purpose of effecting this change, and to fulfil other functions less understood, but no doubt equally important. Diastase is found in the seeds only during the period when the starch they contain is passing into sugar; as soon as that change has taken place, its function is ended, and it disappears.

CHAPTER III.

THE CHANGES WHICH TAKE PLACE IN THE FOOD OF PLANTS DURING THEIR GROWTH.

The simple compounds which the plant absorbs from the atmosphere and soil are elaborated within its system, and converted into the various complex substances of which its tissues are composed, by a series of changes, the details of which are still in some respects imperfectly known, although their general nature is sufficiently well understood. They may be best rendered intelligible by reference, in the first instance, to the changes occurring during germination, when the young plant is nourished by a supply of food stored up in the seed, in sufficient quantity to maintain its existence until the organs by which it is afterwards to draw its nutriment from the air and soil are sufficiently developed to serve that purpose.

Changes occurring during Germination.--When a seed is placed in the soil under favourable circumstances, it becomes the seat of an important and remarkable series of chemical changes, which result in the production of the young plant. Experiment and observation have shown that heat, moisture, and air, are necessary to the production of these changes, and though probably not absolutely essential, the absence of light is favourable in the early stages. The temperature required for germination varies greatly in different seeds, some germinating readily at a few degrees above the freezing point, and others requiring a tolerably high temperature. The rapidity with which it takes place appears to increase with the temperature; but this is true only within very narrow limits, for beyond a certain point heat is injurious, and when it exceeds 120?or 130?Fahrenheit, entirely prevents the process. The presence of oxygen is also essential, for it has been shown that if seeds are placed in a soil exposed to an atmosphere deprived of that element, or if they be buried so deep that the air does not reach them, they may lie without change for an unlimited period; but so soon as they are exposed to the air, germination immediately commences. Illustrations of this fact are frequently observed where earth from a considerable depth has been thrown up to the surface, when it often becomes covered with plants not usually seen in the neighbourhood, which have sprung from buried seeds. When all the necessary conditions for germination are fulfilled, the seed absorbs moisture, swells up, and sends out a shoot which rises to the surface, and a radicle which descends--the one destined to develop the leaves, the other the roots, by which the plant is afterwards to derive its nutriment from the air and the soil. But until these organs are properly developed, the plant is dependent on the matters contained in the seed itself. These substances are mostly insoluble, but are brought into solution by the atmospheric oxygen acting upon the gluten, and converting it into a soluble substance called diastase, which in its turn reacts upon the starch, converting it first into dextrine, and then into cellulose, and the latter is finally deposited in the form of organised cells, and produces the first little shoot of the plant. At the first moment of germination, the oxygen absorbed appears simply to oxidize the constituents of the seed, but this condition exists only for a very limited period, and is soon followed by the evolution of carbonic acid, water being at the same time formed from the organic constituents of the seed, which gradually diminishes in weight. The amount of this diminution is different with different plants, but always considerable. Boussingault found that the loss of dry substance in the pea amounted in 26 days to 52 per cent, and in wheat to 57 per cent in 51 days. Against this, of course, is to be put the weight of the young plant produced; but this is never sufficient to counterbalance the diminished weight of the seed, for Saussure found that a horse bean and the plant produced from it weighed, after 16 days, less by 29 per cent than the seed before germination. The same phenomenon is observed in the process of malting, which is in fact the artificial germination of barley, the malt produced always weighing considerably less than the grain from which it was obtained. It was believed by Saussure, and the older investigators, that the carbonic acid evolved was entirely produced from starch and sugar; and as these substances may be viewed as compounds of carbon and water, the change was very simply explained by supposing that the carbon was oxidised and converted into carbonic acid and its water eliminated. But this hypothesis is incapable of explaining all the phenomena observed; for woody fibre, which is one of the chief constituents of the young plant, contains more carbon than the starch and sugar from which it must have been produced, and we are, therefore, forced to admit that the action must be more complicated. There is every reason to believe that the nitrogenous constituents of the seed are most abundantly oxidized, for they are remarkably prone to change; but the action of the air is not confined to them, and it appears most probable that all the substances take part in the decomposition, and the process of germination may, in some respects, be compared to decay or putrefaction, which, like it, is attended by the absorption of oxygen and evolution of carbonic acid; but while in the latter case the residual substances remain in a useless state, in the former they at once become part of a new organism.

Changes occurring during the After-growth of the Plant.--When the plant has developed its roots and leaves, and exhausted the store of materials laid up for it in the seed, it begins to derive its subsistence from the surrounding air, and to absorb carbonic acid, water, ammonia, and nitric acid, and to decompose and convert them into the different constituents of its tissues. These changes take place slowly at first, and more rapidly as the organs fitted for the elaboration of its food are developed. The roots and the leaves are equally active in performing this duty, the former absorbing the mineral matters along with the carbonic acid, ammonia, nitric acid, and moisture in the soil, or the manure added to it; the latter gathering the gaseous substances existing in the air. Each of these undergoes a series of changes claiming our consideration.

Decomposition of Carbonic Acid.--Carbonic acid, which appears to be absorbed with equal readiness by the roots, leaves, and stems, undergoes immediate decomposition, its carbon being retained, and its oxygen, in whole or in part, evolved into the air. This decomposition occurs only under the action of the sun's rays, and has been found to be proportionate to the amount of light to which the plant is exposed. It takes place only in the green parts of plants, for though the roots absorb carbonic acid, they cannot decompose it, or evolve oxygen; and the coloured parts, the flowers, fruits, etc., have an entirely opposite effect, absorbing oxygen and giving off carbonic acid. The absorption of carbonic acid and escape of oxygen has been proved by numerous direct experiments by Saussure and others, in which both atmospheric air and artificial mixtures containing an increased quantity of carbonic acid have been employed. Saussure allowed seven plants of periwinkle (Vinca minor) to vegetate in an atmosphere containing 7? per cent of carbonic acid for six days, during each of which the apparatus was exposed for six hours to the sun's rays. The air was analysed both before and after the experiment, and the results obtained were--

Volume Carbonic of the air. Nitrogen. Oxygen. Acid. Before the experiment, 5746 4199 1116 431 After " 5746 4338 1408 0 ---- ---- ---- ---- Difference, 0 +139 +292 -431

In this experiment the whole of the carbonic acid, amounting to 431 volumes, was absorbed, but only 292 volumes of oxygen were given off. Had the carbonic acid been entirely decomposed, and all its oxygen eliminated, its volume would have been equal to that of the acid, or 431, so that in this instance 139 volumes of the oxygen of the carbonic acid have been retained to form part of the tissues of the plant. On the other hand, the nitrogen is found to be increased after the experiment. It might be supposed that the nitrogen evolved had been derived from the decomposition of the nitrogenous constituents of the plant, but this cannot be the true explanation, because in this particular case it greatly exceeded the whole nitrogen contained in the plants experimented on. Its source is not well understood, but Boussingault supposes it to have existed in the interstices of the plant, and to have escaped during the course of the experiment. Saussure found that the oak, the horse-chesnut, and other plants, absorb oxygen and give off carbonic acid in less volumes than the oxygen, while the house-leek and the cactus absorb oxygen without evolving carbonic acid. The absorption and decomposition of carbonic acid takes place only during the day, and matters are entirely reversed during the night, when oxygen is absorbed and carbonic acid eliminated from all parts of the plants.

Although the action occurring during the night is the reverse of that which takes place during the day, it is in no degree to be attributed to a re-oxidation of the carbon which had been deposited in the tissues of the plant. It appears, on the contrary, to be a purely mechanical, and not a chemical process. During the night the sap continues to circulate through the vessels of the plant, and moisture, carrying with it carbonic acid in solution, is absorbed by the roots; but when it reaches the leaves, where the sun's light would have caused its decomposition during the day, it is again exhaled unchanged. The oxygen absorbed during the night must, however, take part in some chemical processes, for if it were merely mechanical, the absorption would not be confined to that gas alone, but would be participated in by the other constituents of the air. Moreover, the amount of absorption varies greatly in different plants--being scarcely appreciable in some, and very abundant in others. Plants containing volatile oils, which are readily converted into resins by the action of oxygen, or those containing tannin or other readily oxidizable substances, take up the largest quantity. This is remarkably illustrated by an experiment in which the leaves of the Agave americana, after twenty-four hours' exposure in the dark, were found to have absorbed only 0? of their volume of oxygen, while those of the fir, in which volatile oil is abundant, had taken up twice, and those of the oak, containing tannin, eighteen times as much oxygen.

In the flowers, both by day and night, there is a constant absorption of oxygen, and evolution of carbonic acid. In fact, an active oxidation is going on, attended by the evolution of heat, which, in the Arum maculatum and some other plants, is so great as to raise the temperature of the flower 10?or 12?above that of the surrounding air.

Decomposition of Water in the Plant.--In addition to the function which water performs in the plant, as the solvent of the different substances which form its nutriment, and hence as the medium through which they pass into its organs, it serves also as a direct food, undergoing decomposition, and yielding hydrogen to the organic substances. Its constituents, along with those of the carbonic acid absorbed, undergo a variety of transformations, and form the principal part of the non-nitrogenous constituents. It has been already observed that starch, sugar, and the other allied substances, may be considered as compounds of carbon with water; and they might be supposed to owe their origin to the carbonic acid losing the whole of its oxygen, and direct combination then ensuing between the residual carbon and a certain proportion of water; but this would imply that the latter substance undergoes no decomposition, and though undoubtedly the simplest view of the case, it is by no means the most probable. It is much more likely that the carbonic acid is only partially decomposed, half its oxygen being separated, and replaced by hydrogen, produced by the decomposition of a certain quantity of water into its elements. Thus, for instance, sugar may be produced from twelve equivalents of carbonic acid and twelve equivalents of water, twenty-four equivalents of oxygen being eliminated, as thus represented:

12 equivalents of carbonic acid, $C_{12}O_{12}O_{12}$ 12 " water, $H_{12}O_{12}$ 1 " sugar, and 24 of ox. $C_{12}H_{12}O_{12} + O_{24}$

It must not be supposed that we are in a condition to assert that sugar is really produced in the manner here shown, the illustration being given merely for the purpose of pointing out how it may be supposed to occur, and on a similar principle it is possible to explain the formation of most other vegetable compounds; and this subject has been very fully discussed by the late Dr. Gregory, in his "Handbook of Organic Chemistry." That water must be decomposed, is evident from the fact, established by analysis, that the hydrogen of the plant generally exceeds the quantity required to form water with its oxygen, so that this excess at least must be produced by the decomposition of water. The hydrogen of the volatile oils, many of which contain no oxygen, and that of the fats, which contain only a small quantity, must manifestly be obtained in a similar manner.

Decomposition of Ammonia.--The nitrogenous or albuminous compounds of vegetables must necessarily obtain their nitrogen from the decomposition either of ammonia or nitric acid, experiment having distinctly shown that they are incapable of absorbing it in the free state from the atmosphere. It has been clearly ascertained that the albuminous substances do not contain ammonia, and it is hence apparent that a complete decomposition of that substance must take place in the plant. No doubt carbonic acid and water take part with it in these changes, which must be of a very complex character, and in the present state of our knowledge it seems hopeless to attempt any explanation of them.

Decomposition of Nitric Acid.--Chemists are not entirely at one as to whether nitric acid is directly absorbed by the plant, or is first converted into ammonia. But there are certain facts connected with the chemistry of the soil, to be afterwards referred to, which seem to us to leave no doubt that it may be directly absorbed; and in that case it must be decomposed, its oxygen being eliminated, and the nitrogen taking part with carbon and hydrogen in the formation of the organic compounds. It must be clearly understood that while such changes as those described manifestly must take place, the explanations of them which have been attempted by various chemists are not to be accepted as determinately established facts; they are at present no more than hypothetical views which have been expressed chiefly with the intention of presenting some definite idea to the mind, and are unsupported by absolute proof; they are only inferences drawn from the general bearings of known facts, and not facts themselves. Although, therefore, they are to be received with caution, they have advantages in so far as they present the matter to us in a somewhat more tangible form than the vague general statements which are all that could otherwise be made.

CHAPTER IV.

THE INORGANIC CONSTITUENTS OF PLANTS.

When treating of the general constituents of plants, it has been already stated that the older chemists and vegetable physiologists, misled by the small quantity of ash found in them, entertained the opinion that mineral matters were purely fortuitous components of vegetables, and were present merely because they had been dissolved and absorbed along with the humus, which was then supposed to enter the roots in solution, and to form the chief food of the plant. This supposition, which could only be sustained at a time when analysis was imperfect, has been long since disproved and abandoned, and it has been distinctly shown by repeated experiment that not only are these inorganic substances necessary to the plant, but that every one of them, however small its quantity, must be present if it is to grow luxuriantly and arrive at a healthy maturity. The experiments of Prince Salm Horstmar, before alluded to, have established beyond a doubt, that while a seed may germinate, and even grow, to a certain extent, in absence of one or more of the constituents of its ash, it remains sickly and stunted, and is incapable of producing either flower or seed.

Of late years the analysis of the ash of different plants has formed the subject of a large number of laborious investigations, by which our knowledge of this subject has been greatly extended. From these it appears that the quantity of ash contained in each plant or part of a plant is tolerably uniform, differing only within comparatively narrow limits, and that there is a special proportion belonging to each individual organ of the plant. This fact may be best rendered obvious by the subjoined table, showing the quantity of ash contained in a hundred parts of the different substances dried at 212? Most of these numbers are the mean of several experiments:--

Table showing the quantity of inorganic matters in 100 parts of different plants dried at 212?

SEEDS.

Wheat 1?7 Barley 2?8 Oats (with husk) 3?0 Oats (without husk) 2?6 Rye 2?0 Millet 3?0 Rice 0?7 Maize 1?0 Peas 2?8 Beans 3?2 Kidney Beans 4?9 Lentils 2?1 Tares 2?0 Buckwheat 2?3 Linseed 4?0 Hemp seed 5?0 Rape seed 4?5

Indian Rape-seed[A] 4?6 Sunflower 3?6 Cotton seed 5?3 Guinea Corn 1?9 Gold of Pleasure 4?0 White Mustard 4?5 Black Mustard 4?1 Poppy 6?6 Niger seed (Guizotia oleifera) 7?0 Earth nut 3?8 Sweet Almond 4?0 Horse-chesnut 2?1 Grape 2?6 Clover 6?9 Turnip 3?8 Carrot 10?3 Sainfoin 5?7 Italian Ryegrass 6?1 Mangold-Wurzel 6?8

STRAWS AND STEMS.

Wheat 4?4 Barley 4?9 Oat 7?4 Winter Rye 5?5 Summer Rye 5?8 Millet 8?2 Maize 3?0 Pea 4?1 Bean 6?9 Tares 6?0 Lentil 5?8 Buckwheat 4?0 Hops 4?2 Flax straw 4?5 Hemp 4?4 Gold of Pleasure 6?5 Rape 4?1 Potato 14?0 Jerusalem Artichoke 4?0

ENTIRE PLANT.

Potato 17?0 Spurry 10?6 Red Clover 8?9 White Clover 8?2 Yellow Clover 8?6 Crimson Clover (T. incarnatum) 10?1 Cow Grass (T. medium) 11?1 Sainfoin 6?1 Ryegrass 6?2 Meadow Foxtail (Alopecurus pratensis) 7?1 Sweet-scented Vernal Grass (Anthoxanthum odoratum) 6?2 Downy Oat Grass (Avena pubescens) 5?2 Bromus erectus 5?1 Bromus mollis 5?2 Cynosurus cristatus 6?8 Dactylis glomeratus 5?1 Festuca duriuscula 5?2 Holcus lanatus 6?7 Hordeum pratense 5?7 Lolium perenne 7?4 Poa annua 2?3 Poa pratensis 5?4 Poa trivialis 8?3 Phleum pratense 5?9 Plantago lanceolata 8?8 Poterium Sanguisorba 7?7 Yarrow 13?5 Rape Kale 8?0 Cow Cabbage 10?0 Asparagus 6?0 Parsley 1?0 Furze 3?1 Chamomile (Anthemis arvensis) 9?6 Wild Chamomile (Matricaria Chamomilla) 9?0 Corn Cockle (Agrostemma Githago) 13?0 Corn Blue Bottle (Centaurea Cyanus) 7?2 Foxglove 10?9 Hemlock (Conium maculatum) 12?0 Sweet Rush (Acorus Calamus) 6?0 Common Reed (Arundo Phragmites) 1?4 Celandine (Chelidonium majus) 6?5 Equisetum fluviatile 23?0 Equisetum hyemale 11?0 " arvense 13?0 " linosum 15?0 Fucus nodosus 19?3 Fucus vesiculosus 27?3 Laminaria digitata 39?8

LEAVES.

Turnip 9?7 Beet 20?0 Kohl-rabi 18?4 Carrot 10?5 Jerusalem Artichoke 28?0 Hemp 22?0 Hop 17?5 Tobacco 22?2 Spinach 19?6 Chicory 15?7 Poplar 23?0 Red Beech 6?0 White Beech 10?1 Oak 9?0 Elm 16?3 Horse-chesnut 9?8 Maple 28?5 Ash 14?6 Fir 2?1 Acacia 18?0 Olive 6?5 Orange 13?3 Potato 15?0

Tussac Grass 7?5

ROOTS AND TUBERS.

Potato 4?6 Jerusalem Artichoke 5?8 Turnip 13?4 Beet 8?7 Kohl-rabi 6?8 Rutabaga 7?4 Carrot 5?0 Belgian White Carrot 6?2 Mangold-Wurzel 8?8 Parsnip 5?2 Radish 7?5 Chicory 5?1 Madder 8?3

WOODS.

Beech 0?8 Apple 1?9 Cherry 0?8 Birch 1?0 Oak 2?0 Walnut 1?7 Lime 5?0 Horse-chesnut 1?5 Olive 0?8 Mahogany 0?1 Vine 2?7 Larch 0?2 Fir 0?4 Scotch Fir 0?7 Filbert 0?0 Chesnut 3?0 Poplar 0?0 Hazel 0?0 Orange 2?4 Vine 2?7

BARKS.

Beech 6?2 Cherry 10?7 Fir 1?9 Oak 6?0 Horse-chesnut 7?5 Filbert 6?0 Cork 1?2

FRUITS.

Plum 0?0 Cherry 0?3 Strawberry 0?1 Pear 0?1 Apple 0?7 Chesnut 0?9 Cucumber 0?3 Vegetable Marrow 5?0

On examining this table it may be observed that, notwithstanding the very great variety in the proportion of ash in different plants, some general relations may be traced. A certain similarity may be observed between those belonging to the same natural family, the seeds of all the cereal grains, for instance, containing in round numbers two per cent of inorganic matters. Leguminous seeds (peas and beans) contain about three per cent, while in rape-seed, linseed, and the other oily seeds, it reaches four per cent. In the stems and straws less uniformity exists, but with the exception of a few extreme cases, the quantity of ash in general approaches pretty closely to five per cent. Still more diversified results are obtained from the entire plants; but this diversity is probably much more apparent than real, and must be, in part at least, dependent on the proportion existing between the stem and leaves, for the leaves are peculiarly rich in ash, and a leafy plant must necessarily yield a higher total percentage of ash, although, if stems and leaves were separately examined, they might not show so conspicuous a difference.

The leaves surpass all other parts of plants, in the proportion of inorganic constituents they contain, the table showing that in some instances, as in the maple and Jerusalem artichoke, they exceed one-fourth of the whole weight of the dry matter. In other leaves, and more especially in those of the conifer? the proportion is much smaller. Taking the average of all the analyses hitherto made, it appears that leaves contain about thirteen per cent of ash, but the variations on either side are so large that little value is to be attached to it except as an indication of the general abundance of mineral matters.

In roots and tubers the variations are less, and all, except the potato and the turnip, contain about seven per cent of ash.

The smallest proportion of mineral matter is found in wood. In one case only does the proportion reach five per cent, while the average scarcely exceeds one, and in the fir the quantity amounts to no more than one six-hundredth of the dry matter. In the bark the quantity is much larger, and may be stated at seven per cent.

The general proportion of ash found in different parts of plants is given in round numbers in the subjoined table:--

Wood 1 Seeds 3 Stems and straws 5 Roots and tubers 7 Bark 7 Leaves 13

The differences in the quantity of ash contained in different parts of plants are obviously intended to serve a useful purpose, and it is interesting to observe that the wood which is destined to remain for a long period, sometimes for several centuries, a part of the plant, contains the smallest proportion, and it is not improbable that what it does contain is really due, not to the actual woody matter itself, but to the sap which permeates its vessels. By this arrangement but a small proportion of these important mineral matters, which the soil supplies in very limited quantity, is locked up within the plant, and those which are absorbed, after circulating through it, and fulfilling their allotted functions, are accumulated in the leaves, and annually returned to the soil.

The different proportions of mineral matters contained in the individual organs of plants is most strikingly illustrated when parallel experiments are made on the same species; but the number of instances in which a sufficiently extensive series of analyses has been made to show this, is comparatively limited, and is confined to the oat, the orange-tree, and the horse chesnut--each of which has formed the subject of a very elaborate investigation. The following table gives the results obtained on the oat:--

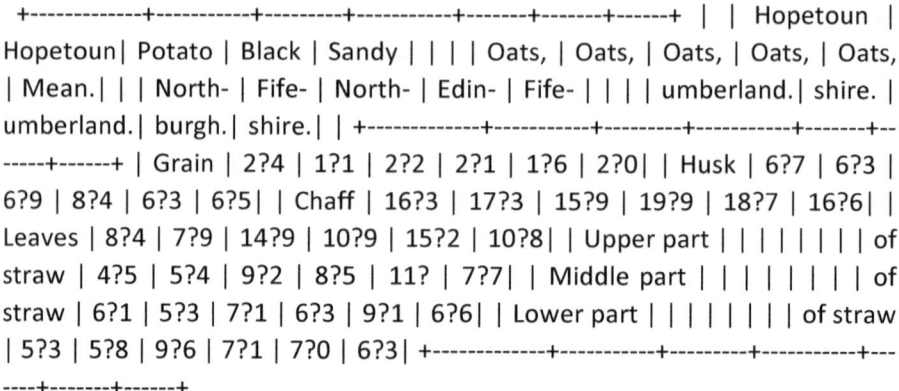

	Hopetoun Mean.	Hopetoun Potato Oats.	Black Oats, North-umberland.	Sandy Oats, Fife-shire.	Oats, North-umberland.	Oats, Edin-burgh.	Oats, Fife-shire.
Grain	2?4	1?1	2?2	2?1	1?6	2?0	
Husk	6?7	6?3	6?9	8?4	6?3	6?5	
Chaff	16?3	17?3	15?9	19?9	18?7	16?6	
Leaves	8?4	7?9	14?9	10?9	15?2	10?8	
Upper part of straw	4?5	5?4	9?2	8?5	11?	7?7	
Middle part of straw	6?1	5?3	7?1	6?3	9?1	6?6	
Lower part of straw	5?3	5?8	9?6	7?1	7?0	6?3	

The specimens of oats on which these analyses were made were from different districts of country, grown on soils of different quality, and were, further, of different varieties; and yet they show, on the whole, a remarkable similarity in the proportion of ash in each part, and indicate that there is a normal quantity belonging to it. Such a series of analyses also affords the most convincing proof that the inorganic matters cannot be fortuitous, and merely absorbed from the soil along with their organic food, as the old chemists supposed, because, in that case, they ought to be uniformly distributed throughout the entire plant, and not accumulated in particular proportions in each individual organ.

Not only does the proportion of ash vary in the different parts of a plant, but even in the same part it is greatly influenced by its period of growth. The laws which regulate these variations are very imperfectly known, but in general it is observed that during the period of active growth the quantity of ash is largest. Thus, it has been found that in early spring the wood of the young shoots of the horse-chesnut contains 9? per cent of ash. In autumn this has diminished to 3?, and the last year's twigs contain only 1? per cent, while in the old wood the quantity does not exceed 0?. Saussure has also observed that the quantity of ash diminishes in certain plants when the seed has ripened. Thus, he found that the percentages of ash, before flowering, and after seeding, were as follows:--

Before flowering. With ripe seed. Sunflower 14? 9? Wheat 7? 3? Maize 12? 4?

On the other hand, the quantity of ash in the leaves of trees increases considerably in autumn, as shown by this table:--

PER-CENTAGE OF ASH IN May. September. Oak leaves 5? 5? Poplar 6? 9? Hazel 6? 7? Horse-chesnut 7? 8?

In general, the proportion of ash appears to increase as the plant reaches maturity, and this is particularly seen in the oat, of which very complete analyses have been made at different periods of its growth:--

Proportion of Ash in different parts of the Oat at different periods of its growth.

Date.	Stalks.	Leaves.	Chaff.	Grain with husk.	Date.	Stalks.	Leaves.	Chaff.	Grain with husk.
2d July	7?3	11?5	...	4?1	9th July	7?0	12?0	...	4?6
16th July	7?4	12?1	6?0	3?8	23d July	7?9	16?5	9?1	3?2
30th July	7?5	16?4	12?8	4?2	5th August	7?3	16?5	13?5	4?1
13th August	6?2	20?7	18?8	4?7	20th August	6?6	21?4	21?7	3?4
27th August	7?1	22?3	22?6	3?1	3d September	8?5	20?0	27?7	3?5

The increase is here principally confined to the leaves and chaff, while the stalks, which owe their strength to a considerable extent to the inorganic matters they contain, are equally supplied at all periods of their growth. In the grain only is there a diminution, but this is apparent and not real, and is due to the fact that the determination of the quantity of ash, as made on the grain with its husk, and the former, which contains only a small quantity of mineral matters, increases much more rapidly in weight than the latter, when it approaches the period of ripening, and it is accordingly during the last three weeks of its growth that this diminution becomes apparent.

The nature of the soil has also a very important influence on the proportion of mineral matters, and of this an interesting illustration is given in the following table, which shows the quantities found in the grain and straw of the same variety of the pea grown on fourteen different soils:--

	Seed.	Straw.
1	2?0	
2	3?5	3?3
3	4?7	3?2
4	3?0	3?9
5	2?9	3?0
6	3?9	6?0
7	2?3	3?0
8	2?7	6?9
9	2?9	3?9
10	1?1	3?1
11	3?1	5?8
12	3?4	7?7
13	2?8	3?6
14	3?1	3?8

Although those differences are very large, especially in the straw, and must be attributed to the soil, it has hitherto been found impossible to ascertain the nature of the relation subsisting between it and the crops it yields; indeed, it must obviously be dependent on very complicated questions, which cannot at present be solved, for it may be observed that the increase in the grain does not occur simultaneously with that in the straw, and in several cases a large proportion of ash in the former is associated with an unusually small amount in the latter. A priori, it might be expected that those soils which are especially rich in the more important constituents of the ash should yield a produce containing more than the average quantity, but this is very far from being an invariable occurrence, and not unfrequently the very reverse is the case. In some instances the variations may be traced to the soil, as in the following analyses of the fruit of the horse-chesnut, grown on an ordinary forest soil, and on a rich soil, produced by the disintegration of porphyritic rock, in which the latter yields a much larger quantity of ash:--

	Kernel of seed.	Green husk.	Brown husk.
Forest soil	2?6	4?3	1?0
Porphyry soil	3?6	7?9	2?0

In the majority of instances we fail to establish any connection between the nature of the soil and the plants it yields, chiefly because we are still very deficient in analyses of those grown on uncultivated soils; and on cultivated land it is impossible to draw conclusions, because the nature of the manure exerts an influence quite as great, if not greater, than that of the soil itself.

The relative proportion in which the different mineral matters enter into the composition of the ash varies within very wide limits, as will be apparent from the following table, containing a selection of the best analyses of our common cultivated and a few uncultivated plants.

Table of the Composition of the Ash of different Plants in 100 Parts.

Note.—Alumina and oxide of manganese occur so rarely, that separate columns have not been introduced for them, but their quantity is stated in notes at the end of the table.

	Potash	Soda	Chloride of Potassium	Chloride of Sodium	Lime	Magnesia
Wheat, grain	30.2	3.2	...	...	1.5	13.9
straw	17.8	2.7	...	...	7.2	1.4
chaff	9.4	1.9	...	...	1.8	1.7
Barley, grain	21.4	...	5.5	1.1	1.5	7.6
straw	11.2	...	...	...	2.4	5.9
Oats, grain[B]	20.3	...	1.3	...	10.8	7.2
straw	19.6	1.3	2.1	4.7	7.1	3.9
chaff[C]	6.3	3.3	...	0.4	1.5	0.8
Rye, grain	33.3	0.9	...	...	2.1	12.1
straw	17.0	...	0.0	0.0	9.0	2.0
Maize, grain	28.7	1.4	...	trace	0.7	13.0
stalks and leaves	35.6	...	...	2.9	10.3	5.2
Rice, grain	20.1	2.9	...	...	7.8	4.6
Buckwheat, straw	31.1	...	7.2	4.5	15.1	1.6
Peas (gray), seed	41.0	...	3.2	1.4	4.8	5.8
straw	21.0	4.2	...	...	37.7	7.7
Beans (common field),						
grain	51.2	0.4	...	...	5.0	6.0
straw	32.5	2.7	...	11.4	19.5	2.3
Tare, straw	32.2	...	3.7	4.3	20.8	5.1
straw	31.2	...	7.1	4.5	15.1	1.6

	Phosphoric Acid	Sulphuric Acid	Carbonic Acid	Silica	Oxide of Iron
Wheat, grain	0.1	46.9	...	...	3.9
straw	0.5	2.5	3.9	...	63.9
chaff	0.7	4.1	...	...	81.2
Barley, grain	2.3	28.3	1.1	...	30.8
straw	1.6	7.0	1.9	...	68.0
Oats, grain	3.5	50.4	...	...	4.0
straw	1.9	5.7	3.5	1.6	49.6
chaff	1.8	1.4	9.1	...	72.5
Rye, grain	1.4	39.2	0.7	...	9.2
straw	1.0	3.0	0.0	...	64.0
Maize, grain	0.7	53.9	...	...	1.5
stalks and leaves	2.8	8.9	5.6	2.7	27.8
Rice, grain	2.2	62.3	...	...	

1?7 | | Buckwheat, straw | ... | 10?4 | 4?7 | 20?7 | 3?7 | | Peas (gray), seed | 0?8 | 36?0 | 4?7 | 0?2 | 0?8 | | straw | 1?7 | 4?5 | 8?8 | 12?8 | 3?3 | | Beans (common field), | | | | | | grain | ... | 28?2 | 3?5 | 3?2 | 0?2 | | straw | 0?1 | 0?9 | 1?0 | 25?2 | 2?1 | | Tare, straw | 0?5 | 10?9 | 2?2 | 18?3 | 1?8 | | straw | ... | 10?4 | 4?7 | 20?7 | 3?7 | +----------------------+--------+----------+-----------+----------+---------+

+---------------------+--------+------+-----------+---------+------+----------+ | | Potash.| Soda.| Chloride | Chloride| Lime.| Magnesia.| | | | | of | of | | | | | | | Potassium.| Sodium. | | | +----------------------+--------+------+-----------+---------+---------+ |Flax, seed | 34?7 | 1?9| ... | 0?6 | 8?0| 13?1 | | straw | 21?3 | 3?8| ... | 9?1 | 21?0| 4?0 | |Rape, seed[D] | 16?3 | 0?4| ... | 0?6 | 8?0| 8?0 | | straw[E] | 16?3 | 10?7| ... | 2?3 | 21?1| 2?2 | |Spurry | 26?2 | 1?4| ... | 8?0 | 14?6| 8?8 | |Chicory root | 34?4 | ...| 8?2 | 2?8 | ...| ... | |Red clover | 25?0 | ...| 9?8 | 6?2 | 21?7| 8?7 | |Cow grass, | | | | | | | | Trifolium medium | 22?8 | ...| 12?9 | 1?6 | 24?2| 8?6 | |Yellow clover | 27?8 | ...| 11?2 | 8?6 | 17?6| 8?9 | |Alsike clover | 29?2 | ...| 6?9 | 1?5 | 26?3| 4?1 | |Lucerne | 27?6 | ...| 11?4 | 1?1 | 20?0| 5?2 | |Anthoxanthum odoratum| 32?3 | ...| 7?3 | 4?0 | 9?1| 2?3 | |Alopecurus pratensis | 37?3 | ...| 9?0 | ... | 3?0| 1?8 | |Avena pubescens | 31?1 | ...| 4?5 | 5?6 | 4?2| 3?7 | |Bromus erectus | 20?3 | ...| 10?3 | 1?8 | 10?8| 4?9 | |Bromus mollis | 30?9 | 0?3| ... | 3?1 | 6?4| 2?0 | |Cynosurus cristatus | 24?9 | ...| 11?0 | ... | 10?6| 2?3 | |Dactylis glomerata | 29?2 | ...| 17?6 | 3?9 | 5?2| 2?2 | |Festuca duriuscula | 31?4 | ...| 8?7 | 0?2 | 10?1| 2?3 | |Holcus lanatus | 34?3 | ...| 3?1 | 6?6 | 8?1| 3?1 | |Lolium perenne | 24?7 | ...| 13?0 | 7?5 | 9?4| 2?5 | |Annual ryegrass | 28?9 | 0?7| ... | 5?1 | 6?2| 2?9 | |Poa annua | 41?6 | ...| 0?7 | 3?5 | 11?9| 2?4 | |Poa pratensis | 31?7 | ...| 11?5 | 1?1 | 5?3| 2?1 | |Poa trivialis | 29?0 | ...| 6?0 | ... | 8?0| 3?2 | |Phleum pratense | 31?9 | ...| 0?0 | 3?4 | 14?4| 5?0 | |Plantago lanceolata | 33?6 | ... | 4?3 | 8?0 | 19?1| 3?1 | |Poterium Sanguisorba | 30?6 | ... | 3?7 | 1?5 | 24?2| 4?1 | |Achillea Millefolia | 30?7 | ... | 20?9 | 3?3 | 13?0| 3?1 | |Potato, tuber | 43?8 | 0?9| ... | 7?2 | 1?0| 3?7 | | stem | 39?3 | 3?5| ... | 20?3 | 14?5| 4?0 | | leaves | 17?7 | ... | 4?5 | 11?7 | 27?9| 7?8 | |Jerusalem Artichoke | 55?9 | ... | 4?8 | ... | 3?4| 1?0 | | stem | 38?0 | 0?9| ... | 4?8 | 20?1| 1?1 | | leaves | 6?1 | 3?2| ... | 1?2 | 40?5| 1?5 | |Turnip, seed | 21?1 | 1?3| ... | ... | 17?0| 8?4 | | bulb | 23?0 | 14?5| ... | 7?5 | 11?2| 3?8 | | leaves | 11?6 | 12?3| ... | 12?1 | 28?9| 2?2 | |Mangold Wurzel, root | 21?8 | 3?3| ... | 49?1 | 1?0| 1?9 | | leaves | 8?4 | 12?1| ... | 37?6 | 8?2| 9?4 | |Carrot, root | 42?3 | 12?1| ... | ...

	5?4	2?9			
leaves	17?0	4?5	...	3?2	24?5	0?9
Kohl-rabi, bulb	36?7	2?4	...	11?0	10?0	2?6
leaves	9?1	...	5?9	6?6	30?1	3?2
Cow cabbage, head	40?6	2?3	...	...	15?1	2?9
stalk	40?3	4?5	...	2?8	10?1	3?5
Poppy seed	9?0	...	7?5	1?4	35?6	9?9
leaves	36?7	...	2?0	2?1	30?4	6?7
Mustard seed (white)	25?8	0?3	...	...	19?0	5?0
Radish root	21?6	...	1?9	7?7	8?8	3?3
Tobacco leaves	36?7	...	2?0	2?1	30?4	6?7
Fucus nodosus[F]	20?3	4?8	...	24?3	9?0	6?5
Fucus vesiculosus[G]	20?5	6?9	...	24?1	8?2	5?3
Laminaria digitata[H]	12?6	...	2?0	19?4	4?2	10?4

	Phosphoric Acid	Sulphuric Acid	Carbonic Acid	Silica	Oxide of Iron
Flax, seed	0?0	38?4	1?6	0?2	1?5
straw	5?8	7?3	3?9	15?5	7?2
Rape, seed	1?9	31?0	5?8	5?4	19?8
straw	1?0	4?8	3?0	23?4	11?0
Spurry	...	10?0	1?9	27?8	1?4
Chicory root	...	...	...	...	...
Red clover	1?6	4?9	2?6	18?5	1?5
Cow grass, Trifolium medium	1?9	4?4	2?6	20?6	1?2
Yellow clover	1?0	...	4?2	4?1	1?6
Alsike clover	0?1	5?4	3?5	20?4	1?3
Lucerne	2?3	6?7	4?0	15?4	2?3
Anthoxanthum odoratum	1?8	10?9	3?9	1?6	28?5
Alopecurus pratensis	0?7	6?5	2?6	0?5	38?5
Avena pubescens	0?2	10?2	3?7	...	36?8
Bromus erectus	0?6	7?3	5?6	0?5	38?8
Bromus mollis	0?8	9?2	4?1	9?7	33?4
Cynosurus cristatus	0?8	7?4	3?0	...	40?1
Dactylis glomerata	0?9	8?0	3?2	2?9	26?5
Festuca duriuscula	0?8	12?7	3?5	1?8	28?3
Holcus lanatus	0?1	8?2	4?1	1?2	28?1
Lolium perenne	0?1	8?3	5?0	0?9	27?3
Annual ryegrass	0?8	10?7	3?5	...	41?9
Poa annua	1?7	9?1	10?8	3?9	16?3
Poa pratensis	0?8	10?2	4?6	0?0	32?3
Poa trivialis	0?9	9?3	4?7	0?9	37?0
Phleum pratense	0?7	11?9	4?6	4?2	31?9
Plantago lanceolata	0?0	7?8	6?1	14?0	2?7
Poterium Sanguisorba	0?6	7?1	4?4	21?2	0?3
Achillea Millefolia	0?1	7?3	2?4	9?6	9?2
Potato, tuber	0?4	8?1	15?4	18?9	1?4
stem	1?4	6?8	6?6	...	2?6
leaves	4?0	13?0	6?7	...	6?7
Jerusalem Artichoke	0?5	16?9	3?7	11?0	1?2
stem	0?8	2?7	3?3	25?0	1?1
leaves	1?4	6?1	2?1	24?1	17?5
Turnip, seed	1?5	40?7	7?0	0?2	0?7
bulb	0?7	9?1	16?3	10?4	2?9
leaves	3?2	4?5	10?6	6?8	8?4
Mangold Wurzel, root					

| 0?2 | 1?5 | 3?4 | 15?3 | 1?0 | | leaves | 1?6 | 5?9 | 6?4 | 6?2 | 2?5 | |Carrot, root | 0?1 | 12?1 | 4?6 | 18?0 | 1?1 | | leaves | 3?3 | 6?1 | 5?8 | 23?5 | 11?1 | |Kohl-rabi, bulb | 0?8 | 13?5 | 11?3 | 10?4 | 0?3 | | leaves | 5?0 | 9?3 | 10?3 | 8?7 | 9?7 | |Cow cabbage, head | 0?7 | 12?3 | 7?7 | 16?8 | 1?6 | | stalk | 0?1 | 19?7 | 11?1 | 6?3 | 1?4 | |Poppy seed | 0?1 | 31?8 | 1?2 | ... | 3?4 | | leaves | 2?4 | 3?8 | 5?9 | ... | 11?0 | |Mustard seed (white) | 0?9 | 44?7 | 2?9 | ... | 1?1 | |Radish root | 1?9 | 41?9 | 7?1 | ... | 8?7 | |Tobacco leaves | 2?8 | 3?4 | 5?9 | ... | 11?0 | |Fucus nodosus | 0?6 | 1?1 | 21?7 | 6?9 | 0?8 | |Fucus vesiculosus | 0?5 | 2?4 | 28?1 | 2?0 | 0?7 | |Laminaria digitata | 0?5 | 1?5 | 7?6 | 15?3 | 1?0 | +---------------------+--------+----------+-----------+----------+---------+

A simple inspection of this table leads to various interesting conclusions. It is particularly to be observed that some of the constituents of the ash are not invariably present, and two at least--namely, alumina and manganese--are found so rarely as to justify the inference that they are not indispensable. Of the other substances, iodine is restricted exclusively to sea-plants, but to them it appears to be essential. Oxide of iron, which occurs only in small quantities, has sometimes been considered fortuitous, but it is almost invariably present, and the experiments of Prince Salm Horstmar leave no doubt that it is essential to the plant. Its function is unknown, but it is an important constituent of the blood of herbivorous animals, and may be present in the plant, less for its own benefit than for that of the animal of which it is destined to become the food.

Soda appears to be a comparatively unimportant constituent of the ash, of which it generally forms but a small proportion, although the instances of its entire absence are rare. In the cruciferous plants (turnip, rape, etc.) it is found abundantly, and to them it appears indispensable, but in most other plants it admits of replacement by potash. It seems probable that where the soil is rich in the latter substance, plants will select that alkali in preference to soda; but as they must have a certain quantity of alkali, the latter may supply the place of the former where it is deficient. Cultivation, probably by enriching the soil in that element, increases the proportion of potash found in the ash of plants, as is remarkably seen in the asparagus, which gave the following quantities of alkalies and chlorine:--

Wild. Cultivated. Potash 18? 50? Soda 16? trace. Chlorine 16? 8?

The soda having almost entirely disappeared in the cultivated plant, while a corresponding increase had taken place in the quantity of potash.

Potash is one of the most important elements of the ash of all plants, rarely forming less than 20, and sometimes more than 50 per cent of its weight. The latter proportion occurs chiefly in the roots and tubers, but it is also abundant in all seeds and in the grasses. The straw, and particularly the chaff of the cereals, and the leaves of most plants, contain it in smaller quantity, although exceptions to this are not unfrequent, one of the most curious being the case of poppy-seed, which contains only about 12 per cent, while the leaves yield upwards of 37 per cent.

The proportion of lime varies within very wide limits, being sometimes as low as 1, and in other plants reaching 40 per cent of their ash. The former proportion occurs in the grains of the cerealia, and the latter in the leaves of some plants, and more especially in the Jerusalem artichoke. The turnip and some of the leguminous plants also contain it abundantly.

Magnesia is generally found in small quantity. It is largest in the grains, amounting in them to about 12 or 13 per cent of the ash, but in other plants it varies from 2 to 4 per cent. Although small in quantity, it is an important substance, and apparently cannot be dispensed with; at least there is no instance known of its entire absence.

Chlorine is by no means an invariable constituent of the ash, although it is generally present, and sometimes in considerable quantity. It is most abundant when the proportion of soda is large, and exists in the ash principally in combination with that base as common salt. The relation between these two elements may be traced more or less distinctly throughout the whole table of analyses, and conspicuously in that of mangold-wurzel, where the common salt amounts to almost exactly one-half of the whole mineral matter. The analyses of the cultivated and uncultivated asparagus also show that a diminution in the soda is accompanied by a reduction in the proportion of chlorine.

Sulphuric Acid is an essential constituent of the ash. But it is to be observed that it is in some instances entirely, and in all partially, a product of the combustion to which the plant has been submitted in order to obtain the ash. It is partly derived from the sulphur contained in the albuminous compounds, which is oxidised and converted into sulphuric acid during the process of burning the organic matter, and remains in the ash. The quantity of sulphuric acid found in the ash is, however, no criterion of that existing in the plant, for a considerable quantity of it escapes during burning. The extent to which this occurs in particular instances is well illustrated by reference to the case of white mustard, which yields an ash containing only 2?9 of sulphuric acid, equivalent to 0? of sulphur; and if calculated on the seed itself, this will amount to no more than 0?39 per cent, while experiments made in another manner prove it to contain about thirty times as much, or more than 1 per cent. For the purpose of determining the total quantity of sulphur which the plants contain in their natural state, it is necessary to oxidise them by means of nitric acid; and from such experiments the following table, showing the total amount of sulphur contained in 100 parts of different plants, dried at 212? has been constructed:--

Poa palustris 0?65 Lolium perenne 0?10 Italian Ryegrass 0?29 Trifolium pratense 0?07 repens 0?99 Lucerne 0?36 Vetch 0?78 Potato tuber 0?82 tops 0?06 Carrot, root 0?92 tops 0?45 Mangold-Wurzel, root 0?58 tops 0?02 Swede, root 0?35 tops 0?58 Rape 0?48

Drumhead Cabbage 0?31 Wheat, grain 0?68 straw 0?45 Barley, grain, 0?53 straw 0?91 Oats, grain 0?03 straw 0?89 Rye, grain 0?51 Beans 0?56 Peas 0?27 Lentils 0?10 Hops 1?63 Gold of Pleasure 0?53 Black Mustard 1?70 White Mustard 1?50

Phosphoric acid, which may be looked upon as the most important mineral constituent of plants, is found to be present in very variable proportions. The straws, stems, and leaves contain it in comparatively small quantity, but in the seeds of all plants it is very abundant. In these of the cereals it constitutes nearly half of their whole mineral components, and it rarely falls below 30 per cent.

Carbonic acid occurs in very variable quantities in the ash. It is of comparatively little importance in itself, and is really produced by the oxidation of part of the carbonaceous matters of the plant; but it has a special interest, in so far as it shows that part of the bases contained in the plant must in its natural state have been in union with organic acids, or combined in some way with the organic constituents of the plant.

Silica is an invariable constituent of the ash, but in most plants occurs but in small quantity. The cereals and grasses form an exception to this rule, for in them it is an abundant and important element. It is not, however, uniformly distributed through them, but is accumulated to a large extent in the stem, to the strength and rigidity of which it greatly contributes. The hard shining layer which coats the exterior of straw, and which is still more remarkably seen on the surface of the bamboo, consists chiefly of silica; and in the latter plant this element is sometimes so largely accumulated, that concretions resembling opal, and composed entirely of it, are found loose within its joints. The necessity for a large supply of silica in the stems of other plants does not exist, and in them it rarely exceeds 5 or 6 per cent, but in some leaves it is more abundant.

A knowledge of the composition of the ash of plants is of considerable importance in a practical point of view, and enables us in many instances to explain why some plants will not grow upon particular soils on which others flourish. Thus, for instance, a plant which contains a large quantity of lime, such as the bean or turnip, will not grow in a soil in which that element is deficient, although wheat or barley, which require but little lime, may yield excellent crops. Again, if the soil be deficient in phosphoric acid, those plants only will grow luxuriantly which require but a small quantity of that element, and hence it follows that on such a soil plants cultivated for the sake of their stems, roots, or leaves, in which the quantity of phosphoric acid is small, may yield a good return; while others, cultivated for the sake of their seed, in which the great proportion of that constituent of the ash is accumulated, may yield a very small crop. It is obvious also that even where a soil contains a proper quantity of all its ingredients, the repeated cultivation of a plant which removes a large quantity of any individual element, may, in the course of time, so far reduce the amount of that substance as to render the soil incapable of any longer producing that plant, although, if it be replaced by another which requires but little of the element thus removed, it may again produce an abundant crop. On this principle also, attempts have been made to explain the rotation of crops, which has been supposed to depend on the cultivation in successive years of plants which abstract from the soil preponderating quantities of different mineral matters. But though this has unquestionably a certain influence, we shall afterwards see reason to doubt whether it affords a sufficient explanation of all the observed phenomena.

It may be observed, on examining the table of the percentage and position of the ash, that some plants are especially rich in alkalies, while in others lime or silica preponderate, and it would therefore be the object of the farmer to employ, in succession, crops containing these elements in different proportions. In carrying out this view, attempts have been made to classify different plants under the heads of silica plants, lime plants, and potash plants; and the following table, extracted from Liebig's Agricultural Chemistry, in which the constituents of the ash are grouped under the three heads of salts of potash and soda, lime and magnesia, and silica, gives such a classification as far as it is at present possible:--

	Salts of Potash and Soda.	Salts of Lime and Magnesia.	Silica.
Silica Plants. { Oat straw with seeds	34?0	4?0	62?0
{ Wheat straw	22?0	7?0	61?0
{ Barley straw with seeds	19?0	25?0	55?0
{ Rye straw	18?5	16?2	63?9
{ Good hay	6?0	34?0	60?0
Lime Plants { Tobacco	24?4	67?4	8?0
{ Pea straw	27?2	63?4	7?1
{ Potato plant	4?0	59?0	36?0
{ Meadow Clover	39?0	56?0	4?0
Potash Plants. { Maize straw	72?5	6?0	18?0
{ Turnips	81?0	18?0	--
{ Beet root	88?0	12?0	--
{ Potatoes	85?1	14?9	--
{ Jerusalem Artichoke	84?0	15?0	--

The special application of these facts must be reserved till we come to treat of the rotation of crops.

It is manifest that, as the crops removed from the soil all contain a greater or less amount of inorganic matters, they must be continually undergoing diminution, and at length be completely exhausted unless their quantity is maintained from some external source. In many cases the supply of these substances is so large that ages may elapse before this becomes apparent, but where the quantity is small, a system of reckless cropping may reduce a soil to a state of absolute sterility. A remarkable illustration of this fact is found in the virgin soils of America, from which the early settlers reaped almost unheard-of crops, but, by injudicious cultivation, they were soon exhausted and abandoned, new tracts being brought in and cultivated only to be in their turn abandoned. The knowledge of the composition of the ash of plants assists us in ascertaining how this exhaustion may be avoided, and indicates the mode in which such soils may be preserved in a fertile state.

FOOTNOTES:

[Footnote A: Apparently a species of Sinapis.]

[Footnote B: Oxide of Manganese, 0?2.]

[Footnote C: Oxide of Manganese, 0?2.]

[Footnote D: Alumina, 1?2.]

[Footnote E: Alumina, 0?3.]

[Footnote F: Iodide of Potassium, 0?4; Sulphuret of Sodium, 3?6.]

[Footnote G: Iodide of Potassium, 0?3.]

[Footnote H: Iodide of Potassium, 1?8.]

CHAPTER V.

THE SOIL--ITS CHEMICAL AND PHYSICAL CHARACTERS.

No department of agricultural chemistry is surrounded with greater difficulties and uncertainties than that relating to the properties of the soil. When chemistry began to be applied to agriculture, it was not unnaturally supposed that the examination of the soil would enable us to ascertain with certainty the mode in which it might be most advantageously improved and cultivated, and when, as occasionally happened, analysis revealed the absence of one or more of the essential constituents of the plant in a barren soil, it indicated at once the cause and the cure of the defect. But the expectations naturally formed from the facts then observed have been as yet very partially fulfilled; for, as our knowledge has advanced, it has become apparent that it is only in rare instances that it is possible satisfactorily to connect together the composition and the properties of a soil, and with each advancement in the accuracy and minuteness of our analysis the difficulties have been rather increased than diminished. Although it is occasionally possible to predicate from its composition that a particular soil will be incapable of supporting vegetation, it not unfrequently happens that a fruitful and a barren soil are so similar that it is impossible to distinguish them from one another, and cases even occur in which the barren appears superior to the fertile soil. The cause of this apparently anomalous phenomenon lies in the fact that analysis, however minute, is unable to disclose all the conditions of fertility, and that it must be supplemented by an examination of its physical and other chemical properties, which are not indicated by ordinary experiments. Of late years very considerable progress has been made in the investigation of the properties of the soil, and many facts of great importance have been discovered, but we are still unable to assert that all the conditions of fertility are yet known, and the practical application of those recently discovered is still very imperfectly understood.

It must not be supposed that a careful analysis of a soil is without value, for very important practical deductions may often be drawn from it, and when this is not practicable it is not unfrequently due to its being imperfect or incomplete, for it is so complex that the cases in which all the necessary details have been eliminated are even now by no means numerous. In fact, the want of a large number of thorough analyses of soils of different kinds is a matter of some difficulty, and so soon as a satisfactory mode of investigation can be determined upon, a full examination of this subject would be of much importance.

Origin of Soils.--The constituents of the soil, like those of the plant, may be divided into the great classes of organic and inorganic. The origin of the former has been already discussed: they are derived from the decay of plants which have already grown upon the soil, and which, in various stages of decomposition, form the numerous class of substances grouped together under the name of humus. The organic substances may therefore be considered as in a manner secondary constituents of the soil, which have been accumulated in it as the consequence of the growth and decay of successive generations of plants, while the primeval soil consisted of inorganic substances only.

The inorganic constituents of the soil are obtained as the result of a succession of chemical changes going on in the rocks which protrude through the surface of the earth. We have only to examine one of these rocks to observe that it is constantly undergoing a series of important changes. Under the influence of air and moisture, aided by the powerful agency of frost, it is seen to become soft, and gradually to disintegrate, until it is finally converted into an uniform powder, in which the structure of the original rock is with difficulty, if at all distinguishable. The rapidity with which these changes take place is very variable; in the harder rocks, such as granite and mica slate it is so slow as to be scarcely perceptible, while in others, such as the shales of the coal formation, a very few years' exposure is sufficient for the purpose. These actions, operating through a long series of years, are the source of the inorganic constituents of all soils.

Geology points to a period at which the earth's surface must have been altogether devoid of soil, and have consisted entirely of hard crystalline rocks, such as granite and trap, by the disintegration of which, slowly proceeding from the creation down to the present time, all the soils which now cover the surface have been formed. But they have been produced by a succession of very complicated processes; for these disintegrated rocks being washed away in the form of fine mud, or at least of minute particles, and being deposited at the bottom of the primeval seas, have there hardened into what are called sedimentary rocks, which being raised above the surface by volcanic action or other great geological forces, have been again disintegrated to yield different soils. Thus, then, all soils are directly or indirectly derived from the crystalline rocks, those overlying them being formed immediately by their decomposition, while those found above the sedimentary rocks may be traced back through them to the crystalline rocks from which they were originally formed.

Such being the case, the composition of different soils must manifestly depend on that of the crystalline rocks from which they have been derived. Their number is by no means large, and they all consist of mixtures in variable proportions of quartz, felspar, mica, hornblende, augite, and zeolites. With the exception of quartz and augite, these names are, however, representatives of different classes of minerals. There are, for instance, several different minerals commonly classified under the name of felspar, which have been distinguished by mineralogists by the names of orthoclase, albite, oligoclase, and labradorite; and there are at least two sorts of mica, two of hornblende, and many varieties of zeolites.

Quartz consists of pure silica, and when in large masses is one of the most indestructible rocks. It occurs, however, intermixed with other minerals in small crystals, or irregular fragments, and forms the entire mass of pure sand.

The four kinds of felspar which have been already named are compounds of silica with alumina, and another base which is either potash, soda, or lime. Their composition is as follows, two examples of each being given--

	Albite.		Oligoclase.		Labradorite.		Orthoclase.	
Silica	65?2	65?0	67?9	68?3	62?0	63?1	54?6	54?7
Alumina	18?7	18?4	19?1	18?0	23?0	23?9	27?7	27?9
Peroxide of iron	traces	0?3	0?0	1?1	0?2	--	--	0?1
Oxide of manganese	traces	0?3	--	--	--	--	--	--
Lime	0?4	1?3	0?6	1?6	4?0	2?4	12?1	10?0
Magnesia	0?0	1?3	--	0?1	0?2	0?7	--	0?8
Potash	14?2	9?2	--	2?3	1?5	2?9	--	0?9
Soda	1?5	3?9	11?2	7?9	8?0	9?7	5?6	5?5
	100?0	99?7	100?8	99?3	100?9	101?7	100?0	99?9

It is obvious that soils produced by the disintegration of these minerals must differ materially in quality. Those yielded by orthoclase must generally abound in potash, while albite and labradorite, containing little or none of that element, must produce soils in which it is deficient. The quality of the soil they yield is not however entirely dependent on the nature of the particular felspar which yields it, but is also intimately connected with the extent to which the decomposition has advanced. It is observed that different felspars undergo decomposition with different degrees of rapidity but after a certain time they all begin to lose their peculiar lustre, acquire a dull and earthy appearance, and at length fall into a more or less white and soft powder. During this change water is absorbed, and, by the decomposing action of the air, the alkaline silicate is gradually rendered soluble, and at length entirely washed away, leaving a substance which, when mixed with water, becomes plastic, and has all the characters of common clay. The nature of this change will be best seen by the following analysis of the clay produced during this composition, which is employed in the manufacture of porcelain under the name of kaolin, or china clay--

Silica 46?0 Alumina 36?3 Peroxide of iron 3?1 Carbonate of lime 0?5 Potash 0?7 Water 12?4 ---- 100?0

In this instance the decomposition of the felspar had reached its limit, a mere trace of potash being left, but if taken at different stages of the process, variable proportions of that alkali are met with. This decomposition of felspar is the source of the great deposits of clay which are so abundantly distributed over the globe, and it takes place with nearly equal rapidity with potash and soda felspar. It is rarely complete, and the soils produced from it frequently contain a considerable proportion of the undecomposed mineral, which continues for a long period to yield a supply of alkalies to the plants which grow on them.

Mica is a very widely distributed mineral, and two varieties of it are distinguished by mineralogists, one of which is characterised by the large quantity of magnesia it contains. Different specimens are found to vary very greatly in composition, but the following analyses may represent their most usual composition:

MICA. |----------------| Potash. Magnesia. Silica 46?6 42?5 Alumina 36?0 12?6 Peroxide of iron 4?3 --- Protoxide of iron --- 7?1 Oxide of manganese 0?2 1?6 Magnesia --- 25?5 Potash 9?2 6?3 Hydrofluoric acid 0?0 0?2 Water 1?4 3?7 --- - ---- 99?7 99?5

Mica undergoes decomposition with extreme slowness, as is at once illustrated by the fact that its shining scales may frequently be met with entirely unchanged in the soil. Its persistence is dependent on the small quantity of alkaline constituents which it contains; and for this reason it is observed that the magnesian micas undergo decomposition less rapidly than those containing the larger quantity of potash. Eventually, however, both varieties become converted into clay, their magnesia and potash passing gradually into soluble forms.

Hornblende and augite are two widely distributed minerals, which are so similar in composition and properties that they may be considered together.

Of the former two varieties, basaltic and common have been distinguished, and their composition is given below:--

Hornblende. |----------------| Common. Basaltic. Augite.

Silica 41·0 42·4 50·2 Alumina 15·5 13·2 4·0 Protoxide of iron 7·5 14·9 11·0 Oxide of manganese 0·5 0.33 -- Lime 14·9 12·4 20·5 Magnesia 19·0 13·4 13·0 Water 0·0 -- -- ---- ---- ---- 99·4 97·5 99·7

In these minerals alkalies are entirely absent, and their decomposition is due to the presence of protoxide of iron, which readily absorbs oxygen from the air, when the magnesia is separated and a ferruginous clay left.

The minerals just referred to, constitute the great bulk of the mountain masses, but they are associated with many others which take part in the formation of the soil. Of these the most important are the zeolites which do not occur in large masses but are disseminated through the other rocks in small quantity. They form a large class of minerals of which Thomsonite and natrolite may be selected as examples--

Thomsonite. Natrolite.

Silica 38·3 48·8 Alumina 30·4 26·6 Lime 13·3 -- Potash 0·4 0·3 Soda 3·5 16·0 Water 13·9 9·5 ---- ---- 100·8 100·3

They are chiefly characterized by containing their silica in a soluble state, and hence may yield that substance to the plants in a condition particularly favourable for absorption.

It is obvious from what has been stated that all these minerals are capable, by their decomposition, of yielding soft porous masses having the physical properties of soils, but most of them would be devoid of many essential ingredients, while not one of them would yield either phosphoric acid, sulphuric acid, or chlorine. It has, however, been recently ascertained that certain of these minerals, or at least the rocks formed from them, contain minute, but distinctly appreciable traces of phosphoric acid, although in too small quantity to be detected by ordinary analysis; and small quantities of chlorine and sulphuric acid may also in most instances be found.

Still it will be observed that most of these minerals would yield a soil containing only two or three of those substances, which, as we have already learned, are essential to the plant. Thus, potash felspar, while it would give abundance of potash, would be but an inefficient source of lime and magnesia; and labradorite, which contains abundance of lime, is altogether deficient in magnesia and potash.

Nature has, however, provided against this difficulty, for she has so arranged it that these minerals rarely occur alone, the rocks which form our great mountain masses being composed of intimate mixtures of two or more of them, and that in such a manner that the deficiencies of the one compensate those of the other. We shall shortly mention the composition of these rocks.

Granite is a mixture of quartz, felspar, and mica in variable proportions, and the quality of the soil it yields depends on whether the variety of felspar present be orthoclase or albite. When the former is the constituent, granite yields soils of tolerable fertility, provided their climatic conditions be favourable; but it frequently occurs in high and exposed situations which are unfavourable to the growth of plants. Gneiss is a similar mixture, but characterised by the predominance of mica, and by its banded structure. Owing to the small quantity of felspar which it contains, and the abundance of the difficulty decomposable mica, the soils formed by its disintegration are generally inferior. Mica slate is also a mixture of quartz, felspar, and mica, but consisting almost entirely of the latter ingredient, and consequently presenting an extreme infertility. The position of the granite, gneiss, and mica slate soils in this country is such that very few of them are of much value; but in warm climates they not unfrequently produce abundant crops of grain. Syenite is a rock similar in composition to granite, but having the mica replaced by hornblende, which by its decomposition yields supplies of lime and magnesia more readily than they can be obtained from the less easily disintegrated mica. For this reason soils produced from the syenitic rocks are frequently possessed of considerable fertility.

The series of rocks of which greenstone and trap are types, and which are very widely distributed, differ greatly in composition from those already mentioned. They are divisible into two great classes, which have received the names of diorite and dolerite, the former a mixture of albite and hornblende, the latter of augite and labradorite, sometimes with considerable quantities of a sort of oligoclase containing both soda and lime, and of different kinds of zeolitic minerals. Generally speaking, the soils produced from diorite are superior to those from dolerite. The albite which the former contains undergoes a rapid decomposition, and yields abundance of soda along with some potash, which is seldom altogether wanting, while the hornblende supplies both lime and magnesia. Dolerite, when composed entirely of augite and labradorite, produces rather inferior soils; but when it contains oligoclase and zeolites, and comes under the head of basalt, its disintegration is the source of soils remarkable for their fertility; for these latter substances undergoing rapid decomposition furnish the plants with abundant supplies of alkalies and lime, while the more slowly decomposing hornblende affords the necessary quantity of magnesia. In addition to these, the basaltic rocks are found to contain appreciable quantities of phosphoric acid, so that they are in a condition to yield to the plant almost all its necessary constituents.

The different rocks now mentioned, with a few others of less general distribution, constitute the whole of our great mountain masses; and while their general composition is such as has been stated, they frequently contain disseminated through them quantities of other minerals which, though in trifling quantity, nevertheless add their quota of valuable constituents to the soils. Moreover, the exact composition of the minerals of which the great masses of rocks are composed is liable to some variety. Those which we have taken as illustrations have been selected as typical of the minerals; but it is not uncommon to find albite containing 2 or 3 per cent of potash, labradorite with a considerable proportion of soda, and zeolitic minerals containing several per cent of potash, the presence of which must of course considerably modify the properties of the soils produced from them. They are also greatly affected by the mechanical influences to which the rocks are exposed; and being situated for the most part in elevated positions, they are no sooner disintegrated than they are washed down by the rains. A granite, for instance, as the result of disintegration, has its felspar reduced to an impalpable powder, while its quartz and mica remain, the former entirely, the latter in great part, in the crystalline grains which existed originally in the granite. If such a disintegrated granite remains on the spot, it is easy to see what its composition must be; but if exposed to the action of running water, by which it is washed away from its original site, a process of separation takes place, the heavy grains of quartz are first deposited, then the lighter mica, and lastly the felspar. Thus there may be produced from the same granite, soils of very different nature and composition, from a pure and barren sand to a rich clay formed entirely of felspathic debris.

The sedimentary or stratified rocks are formed of particles carried down by water and deposited at the bottom of the primeval seas from which they have been upheaved in the course of geological changes. The process of their formation may be watched at the present day at the mouths of all great rivers, where a delta composed of the suspended matters carried down by the waters is slowly formed. The nature of these rocks must therefore depend entirely on that of the country through which the river flows. If its course runs through a country in which lime is abundant, calcareous rocks will be deposited, and if it passes through districts of different geological characters the deposit must necessarily consist of a mixture of the disintegrated particles of the different rocks the river has encountered. For this reason it is impossible to enter upon a detailed account of their composition. It is to be observed, however, that the particles of which they are composed, though originally derived from the crystalline rocks, have generally undergone a complex series of changes, geology teaching that, after deposition, they may in their turn undergo disintegration and be carried away by water, to be again deposited. Their composition must therefore vary not merely according to the nature of the rock from which they have been formed, but also according to the extent to which the decomposition has gone, and the successive changes to which they have been exposed. They may be reduced to the three great classes of clays, including the different kinds of clay slates, shales, etc., sandstone and limestone. It must be added also, that many of them contain carbonaceous matters produced by the decomposition of early races of plants and animals, and that mixtures of two or more of the different classes are frequent.

The purest clays are produced by the decomposition of felspar, but almost all the crystalline rocks may produce them by the removal of their alkalies, iron, lime, etc. Where circumstances have been favourable, the whole of these substances are removed, and the clay which remains consists almost entirely of silica and alumina, and yields a soil which is almost barren, not merely on account of the deficiency of many of the necessary elements of plants, but because it is so stiff and impenetrable that the roots find their way into it with difficulty. It rarely happens, however, that decomposition has advanced so far as to remove the whole of the alkalies, which is exemplified by the following analyses of the fire clay of the coal formation, and of transition clay slate:--

Transition Fire Clay. Clay Slate.

Silica 60?3 54?7 Alumina 14?1 28?1 Peroxide of iron 8?4 4?2 Lime 2?8 0?8 Magnesia 4?2 1?4 Potash 3?7 1?0 Soda -- 0?4 Carbonic acid } 5?7 8?4 Water } ---- ---- 99?2 99?0

The sandstones are derived from the siliceous particles of granite and other rocks, and consist in many cases of nearly pure silica, in which case their disintegration produces a barren sand, but they more frequently contain an admixture of clay and micaceous scales, which sometimes form a by no means inconsiderable portion of them. Such sandstones yield soils of better quality, but they are always light and poor. Where they occur interstratified with clays, still better soils are produced, the mutual admixture of the disintegrated rocks affording a substance of intermediate properties, in which the heaviness of the clay is tempered by the lightness of the sandstone.

Limestone is one of the most widely distributed of the stratified rocks, and in different localities occurs of very different composition. Limestones are divided into two classes, common and magnesian; the former a nearly pure carbonate of lime, the latter a mixture of that substance with carbonate of magnesia. But while these are the principal constituents, it is not uncommon to find small quantities of phosphate and sulphate of lime, which, however trifling their proportions, are not unimportant in an agricultural point of view. The following analyses will serve to illustrate the general composition of these two sorts of limestone as they occur in the early geological formations:-

COMMON. MAGNESIAN. |-------------------------| |----------------------| Mid-Lothian. Sutherland. Sutherland. Dumfries.

Silica 2?0 7?2 6?0 2?1 Peroxide of iron } 0?5 0?6 1?7 2?0 and alumina } Carbonate of lime 93?1 84?1 50?1 58?1 Carbonate of } 1?2 7?5 41?2 36?1 magnesia } Phosphate of lime 0?6 ... ... Sulphate of lime 0?2 ... ... 0?0

Organic matter 0?0 ... ... Water 0?0 ... 0?9 ... ---- ---- ---- ---- 99?6 99?4 99?9 99?3

These limestones are hard and possess to a greater or less extent a crystalline texture. They are replaced in later geological periods by others which are much softer, and often purer, of which the oolitic limestones, so called from their resemblance to the roe of a fish, and chalk are the most important. Other limestones are also known which contain an admixture of clay. The soils produced by the disintegration of limestone and chalk are generally light and porous, but when mixed with clay, possess a very high degree of fertility, and this is particularly the case with chalk, which yields some of the most valuable of all soils. But it is true only of the common limestones, for experience has shown that those which contain magnesia in large quantity are often prejudicial to vegetation, and sometimes yield barren or inferior soils.

Such are the general characters of the three great classes of stratified rocks; any attempt to particularise the numerous varieties of each would lead us far beyond the limits of the present work. It is necessary, however, to remark, that in many instances one variety passes into the other, or, more correctly speaking, sedimentary rocks occur, which are mixtures of two or more of the three great classes. In fact, the name given to each really expresses only the preponderating ingredient, and many sandstones contain much clay, shales and clay slates abound in lime, and limestones in sand or clay, so that it may sometimes be a matter of some difficulty to decide to which class they belong. Such mixtures usually produce better soils than either of their constituents separately, and accordingly, in those geological formations in which they occur, the soils are generally of excellent quality. The same effect is produced where numerous thin beds of members of the different classes are interstratified, the disintegrated portions being gradually intermixed, and valuable soils formed.

The fertility of the soils formed from the stratified rocks is also increased by the presence of organic remains which afford a supply of phosphoric acid, and which are sometimes so abundant as to form a by no means unimportant part of their mass. They do not occur in the oldest sedimentary rocks, but as we ascend to the more recent geological epochs, they increase in abundance, until, in the greensands and other recent formations, whole beds of coprolites and other organic remains are met with. Great differences are observed in the quality of the soils yielded by different rocks. In general, those formed by the disintegration of clay slates are cold, heavy, and very difficult and expensive to work; those of sandstone light and poor, and of limestone often poor and thin. These statements must, however, be considered as very general; for individual cases occur in which some of these substances may produce good soils, remarkable exceptions being offered by the lower chalk and some of the shales of the coal formation. Little is at present known regarding the peculiar nature of many of these rocks, or their composition; and the cause of the differences in the fertility of the soil produced from them is a subject worthy of minute investigation.

Chemical Composition of the Soil.--Reference has been already made to the division of the constituents of the soil into the two great classes of organic and inorganic. And when treating of the sources of the organic constituents of plants, we entered with some degree of minuteness into the composition and relations of the different members of the former class, and expressed the opinion that they did not admit of being directly absorbed by the plant. But though the parts then stated lead to the inference that, as a direct source of these substances, humus is unimportant, it has other functions to perform which render it an essential constituent of all fertile soils. These functions are dependent partly on the power which it has of absorbing and entering into chemical composition with ammonia, and with certain of the soluble inorganic substances, and partly on the effect which the carbonic acid produced by its decomposition exerts on the mineral matters of the soil. In the former way, its effects are strikingly seen in the manner in which ammonia is absorbed by peat; for it suffices merely to pour upon some dried peat a small quantity of a dilute solution of ammonia to find its smell immediately disappear. This peculiar absorptive power extends also to the fixed alkalies, potash and soda, as well as to lime and magnesia, and has an important effect in preventing these substances being washed out of the soil--a property which, as we shall afterwards see, is possessed also by the clay contained in greater or less quantity in most soils. On the other hand, the air and moisture which penetrate the soil cause its decomposition, and the carbonic acid so produced attacks the undecomposed minerals existing in it, and liberate the valuable substances they contain.

In considering the composition of a soil, it is important to bear in mind that it is a substance of great complexity, not merely because it contains a large number of chemical elements, but also because it is made up of a mixture of several minerals in a more or less decomposed state. The most cursory examination shows that it almost invariably contains sand and scales of mica, and other substances can often be detected in it. Now it has been already observed that the minerals of which soils are composed, differ to a remarkable extent in the facility with which they undergo decomposition, and the bearing of this fact on its fertility is a matter of the highest importance, for it has been found that the mere presence of an abundant supply of all the essential constituents of plants is not always sufficient to constitute a fertile soil. Two soils, for instance, may be found on analysis to have exactly the same composition, although in practice one proves barren and the other fertile. The cause of this difference lies in the particular state of combination in which the elements are contained in them, and unless this be such that the plant is capable of absorbing them, it is immaterial in what quantity they are present, for they are thus locked up from use, and condemn the soil to hopeless infertility.

It is admitted that unless the substances be present in a state in which they can be dissolved, the plant is incapable of absorbing them; but it is a matter of doubt whether it is necessary that they be actually dissolved in the water which permeates the soil, or whether the plant is capable of exercising a directly solvent action. The latter view is the most probable, but at the same time it cannot be doubted, that if they are presented to the plant in solution, they will be absorbed in that state in preference to any other. Hence it has been considered important in the analysis of a soil, not to rest content with the determination of the quantity of each element it contains, but to obtain some indication of the state of combination in which it exists, so as to have some idea of the ease or difficulty with which they may be absorbed. For this purpose it is necessary to determine, 1st, The substances soluble in water; 2d, Those insoluble in water, but soluble in acids; 3d, Those insoluble both in water and acids; and if to these the organic constituents be added, there are four separate heads under which the components of a soil ought to be classified. This classification is accordingly adopted in the most careful and minute analyses; but the difficulty and labour attending them has hitherto precluded the possibility of making them except in a few instances; and, generally speaking, chemists have been contented with treating the soil with an acid, and determining in the solution all that is dissolved. Such analyses are often useful for practical purposes, as for example, when they show the absence of lime, or any other individual substance, by the addition of which we may rectify the deficiency of the soil; but they are of comparatively little scientific value, and throw but little light on the true constitution of the soil, and the sources of its fertility. Nor is it likely that much satisfactory information will be obtained until the number of minute analyses is so far extended as to establish the fundamental principles on which the various properties of the soil depends.

The separation of the constituents of a soil into the four great groups already mentioned, is effected in the following manner:--A given quantity of the soil is boiled with three or four successive quantities of water, which dissolves out all the soluble matters. These generally amount to about one-half per cent of the whole soil, and consist of nearly equal proportions of organic and inorganic substances. In very light and sandy soils, it occasionally happens that not more than one or two-tenths per cent dissolve in water, and in peaty soils, on the other hand, the proportion is sometimes considerably increased, principally owing to the abundance of soluble organic matters.

When the residue of this operation is treated with dilute hydrochloric acid, the matters soluble in acids are obtained in the fluid. The proportion of these substances is liable to very great variations, and in some soils of excellent quality, and well adapted to the growth of wheat, it does not exceed 3 per cent; while in calcareous soils, such as those of the chalk formation, it may reach as much as 50 or 60 per cent. In general, however, it amounts to about 10 per cent. The organic constituents are also very variable in amount; ordinary soils of good quality containing from 2 to 10 per cent, while in peat soils they not unfrequently reach 30 or even 50 per cent. But these cannot be considered fertile soils. The insoluble constituents are likewise subject to great variations, but, in the ordinary clay and sandy soils of this country, they generally form from 70 to 85 per cent of the whole.

The distribution of the constituents under these different heads will be best illustrated by a few analyses of soils of good quality, and for this purpose we shall select two, noted for the excellent crops of wheat they produce, and for their general fertility. The analyses were made from the upper 10 inches, and a quantity of the 10 inches immediately subjacent was analysed as subsoil.

The first is the ordinary wheat soil of the county of Mid-Lothian, the other the alluvial soil of the Carse of Gowrie in Perthshire, so celebrated for the abundance and luxuriance of the crops it produces.

	Mid-Lothian.		Perthshire.	
	Soil.	Subsoil.	Soil.	Subsoil.
SUBSTANCES SOLUBLE IN WATER.				
Silica	0?149	0?104	0?072	0?461
Lime	0?300	0?072	0?184	0?306
Magnesia	0?097	0?016	0?040	0?034
Chlor. of magnesium	--	--	--	0?033
Potash	0?034	0?037	--	--
Soda	0?065	0?049	--	--
Chloride of potassium	--	--	0?088	0?080
Chloride of sodium	--	--	0?110	0?166
Sulphuric acid	0?193	0?124	0?089	0?239
Chlorine	trace	trace	--	--
Organic matters	0?481	0?228	0?608	0?342
	0?319	0?630	0?191	0?661
SOLUBLE IN ACIDS.				
Silica	0?490	0?680	0?482	0?697
Peroxide of iron	5?730	3?820	4?700	4?633
Alumina	2?540	1?130	2?900	3?070
Lime	0?470	0?810	0?616	0?050
Magnesia	0?120	0?850	0?960	0?420
Potash	0?650	0?650	0?445	0?670
Soda	0?050	0?560	0?242	0?920
Sulphuric acid	0?250	0?850	0?911	0?160
Phosphoric acid	0?300	0?970	0?400	0?680
Carbonic acid	--	--	0?500	--
	8?600	6?320	9?156	10?300
INSOLUBLE IN ACIDS.				
Silica	71?890	82?090	63?400	61?200
Alumina	4?810	3?120	11?500	10?400
Peroxide of iron	trace	trace	--	1?670
Lime	0?520	0?500	0?500	0?400
Magnesia	0?610	0?500	0?200	0?450
Potash	0?860	--	2?500	2?030
Soda	0?220	--	1?100	0?440
	78?910	87?210	79?200	77?590
ORGANIC MATTERS.				
Insoluble organic matter }	8?777	4?370	7?400	6?910
Humine	0?850	0?450	0?700	0?840
Humic acid	0?340	0?310	0?800	0?600
Apocrenic acid	0?533	--	--	0?929
Water	2?840	1?670	2?000	4?750
	12?340	6?800	11?900	11?020
Sum of all the constituents	100?169	100?960	99?447	99?571
AMOUNT OF CARBON, HYDROGEN, NITROGEN, AND OXYGEN CONTAINED IN 100 PARTS OF EACH SOIL.				
Carbon	4?10	1?060	2?5	2?3
Hydrogen	0?50	0?324	0?1	0?3
Nitrogen	0?20	0?973	0?1	0?7
Oxygen	4?18	3?001	5?8	4?9

| |--------|----------|---------|------- | | 10?98 | 4?358 | 8?5 | 6?2 --------------------
----

In examining these analyses, it is particularly worthy of notice that by far the larger proportion of the substances soluble in water consists of organic matter, lime, and sulphuric acid, the two last being in combination as sulphate of lime, while some of those substances which are usually considered to be the most important mineral constituents of plants are present in very small quantity--potash, for instance, forming not more than 1-25,000th of the whole soil, and phosphoric acid being entirely absent. On the other hand, this portion contains the whole of the chlorine which exists in the soil, and this might be anticipated from the ready solubility in water of the compounds of that substance.

The portion soluble in acids consists of alumina and oxide of iron, both of which are comparatively unimportant to the plant, but very important, as we shall afterwards see, in relation to the physical properties of the soil. The remainder of the substances soluble in acids, amounting to from 1 and 2 per cent, is composed of some of the most essential constituents of plants. Lime, magnesia, potash, and soda, appear again in larger quantity than in the soluble part, and along with them we have the phosphoric acid to the amount of from 0? to 0? per cent of the whole soil, and sulphuric acid in much smaller quantity.

The insoluble matters differ remarkably in the two soils, that from the Carse of Gowrie being characterised by a large quantity of potash and soda, indicating an important difference in the materials from which they have been formed. In the Perthshire soil it is obvious that the felspathic element has been abundant, and that its decomposition has been arrested at a time, when it still contained a large quantity of alkalies. And this difference is of great practical importance, because those soils, which contain a large quantity of potash in their insoluble portion, have within them a source of permanent fertility, the alkali being gradually liberated by the decomposition which is constantly in progress, owing to the air and moisture permeating the soil. As regards the special distribution of the inorganic matters, it is to be observed that some of them occur in each of the three heads under which they are arranged, while others are confined to one or two. Silica and the alkalies occur generally, though not invariably, in all three. Chlorine is met with only in the part soluble in water, phosphoric acid only in that soluble in acids, while sulphuric acid occurs in both the last-named divisions.

The greater part of the organic matters are insoluble both in water and acids. At least it is generally believed that any portion dissolved by strong acids, in the course of analysis, has been entirely decomposed, and is in a completely different state from that in which it existed actually in the soil.

As an example of a calcareous soil, forming a striking contrast to those given above, we select one from the island of Antigua, from which very large crops of sugar-cane are obtained. The soil is of great depth, and analyses of the subsoil at the depth of 18 inches and 5 feet are given. These last analyses are not so minute as that of the soil itself, the soluble matters not having been separately determined, but included in that soluble in acids.

	Surface Soil.	18 inches deep.	5 feet deep.
SOLUBLE IN WATER.			
Lime	0·7	...	...
Magnesia	trace	...	...
Potash	0·6	...	...
Soda	0·4	...	...
Chlorine	0·5	...	...
Organic matter	0·5	...	...

	2·7		
SOLUBLE IN ACIDS.			
Silica	0·4	...	...
Peroxide of iron	2·2	1·7	1·7
Protoxide of iron	0·7	9·5	3·0
Alumina	1·0	2·2	4·1
Lime	10·3	3·4	25·5
Magnesia	0·0	0·4	0·1
Potash	0·3	0·9	0·8
Soda	0·2	0·1	0·6
Sulphuric acid	trace	0·2	0·3
Phosphoric acid	0·4	trace	0·4
Carbonic acid	7·8	0·2	20·3
	-----	-----	-----
	23·3	18·6	56·8
INSOLUBLE IN ACIDS.			
Silica	41·4	51·4	27·7
Protoxide of iron	3·4	0·6	1·0
Alumina	9·0	1·0	1·0
Lime	0·8	0·8	trace
Magnesia	0·0	0·4	trace
Potash	...	0·4	...
Soda	...	0·5	...
	-----	-----	-----
	54·6	55·1	30·7
ORGANIC MATTERS.			
Humine	1·8 }		
Humic acid	1·5 }	12·5	7·9
Insoluble organic matters	7·6 }		
Water	11·3	14·9	6·6
	-----	-----	-----
	21·2	26·4	13·5
Sum of all the constituents	100·8	100·1	99·0

In this soil there is a general resemblance in the composition of the portion soluble in water to those of the wheat soils. But the part soluble in acids is distinguished by the great abundance of carbonate of lime.

The subsoil contains also a large quantity of protoxide of iron, a substance frequently found in subsoils containing much organic matter, and to which the air has imperfect access. Under these circumstances peroxide of iron is reduced to protoxide; and when present abundantly in the soil in that form, iron has been found to exercise a very injurious influence on vegetation; and it has frequently happened that when subsoils containing it have been brought up to the surface, they have in the first instance caused a manifest deterioration of the soil, although after some time, when it had become peroxidised by the action of the air, it ceased to be injurious.

The soil of Holland, from the neighbourhood of the Zuider Zee, which is an alluvial deposit from the waters of the Rhine, and produces large crops, gave the results which follow--

	Surface.	15 inches deep.	30 inches deep.
Insoluble silica	57?46	51?06	55?72
Soluble silica	2?40	2?96	2?86
Alumina	1?30	2?00	2?88
Peroxide of iron	9?39	10?05	11?64
Protoxide of iron	0?50	0?63	0?00
Oxide of manganese	0?88	0?54	0?84
Lime	4?92	5?96	2?80
Magnesia	0?30	0?40	0?28
Potash	1?26	1?30	1?21
Soda	1?72	2?69	1?37
Ammonia	0?60	0?78	0?75
Phosphoric acid	0?66	0?24	0?78
Sulphuric acid	0?96	1?04	0?76
Carbonic acid	6?85	6?40	4?75
Chlorine	1?40	1?02	1?18
Humic acid	2?98	3?91	3?28
Crenic acid	0?71	0?31	0?37
Apocrenic acid	0?07	0?60	0?52
Other organic matters and } Combined water }	8?24	7?00	9?48
Loss	0?40	0?11	0?53
	-------	-------	-------
	100?00	100?00	100?00

It is unnecessary to multiply analyses of fertile soils, those now given being sufficient to show their general composition. They are all characterised by the presence, in considerable quantity, of all the essential constituents of plants, in a state in which they may be readily absorbed. The absence of one or more of these substances immediately diminishes or altogether destroys the fertility of the soil; and the extent to which this occurs is illustrated by the following analysis of a soil from Pumpherston, Mid-Lothian, forming a small patch in the lower part of a field, and on which nothing would grow. Being naturally wet, it had been drained and sowed with oats, which died out about six weeks after sowing, and left a bare soil on which weeds did not show the slightest disposition to grow.

SOLUBLE IN ACIDS.

Soluble silica 0?73 Peroxide of iron 6?75 Alumina 1?50 Oxide of manganese trace Carbonate of lime 0?56 Magnesia 0?99 Potash 0?32 Soda 0?23 Phosphoric acid trace Chlorine trace ---- 9?08 Silica 73?96 Peroxide of iron 1?71 Alumina 4?63 Lime 0?58 Magnesia 0?20 ---- 80?08 Organic matter 8?12 Water 2?91 ---- 10?03 ------ 99?19

In this instance the barrenness of the soil is distinctly traceable to the deficiency of phosphoric acid, sulphuric acid, and chlorine. There is also a remarkably large quantity of oxide of iron, which, when acted on by the humic acid, is well known to be highly prejudicial to vegetation, and that this took place was shown by the fact that the drains, a couple of months after being laid, were almost stopped up by humate of iron. Still more striking are the following analyses:--

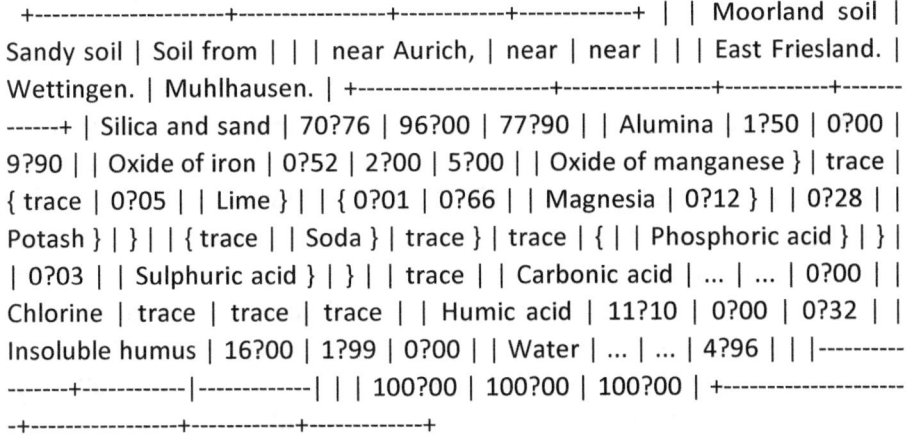

```
+---------------------+-----------------+-----------+-------------+ |   | Moorland soil |
Sandy soil | Soil from |   |   | near Aurich, | near | near |   |   | East Friesland. |
Wettingen. | Muhlhausen. | +---------------------+-----------------+-----------+-------
------+ | Silica and sand | 70?76 | 96?00 | 77?90 |   | Alumina | 1?50 | 0?00 |
9?90 |   | Oxide of iron | 0?52 | 2?00 | 5?00 |   | Oxide of manganese } | trace |
{ trace | 0?05 |   | Lime } |   | { 0?01 | 0?66 |   | Magnesia | 0?12 } |   | 0?28 |   |
Potash } |  } |   | { trace |   | Soda } | trace } | trace | { |   | Phosphoric acid } |   } |
| 0?03 |   | Sulphuric acid } |   } |   | trace |   | Carbonic acid | ... | ... | 0?00 |   |
Chlorine | trace | trace | trace |   | Humic acid | 11?10 | 0?00 | 0?32 |   |
Insoluble humus | 16?00 | 1?99 | 0?00 |   | Water | ... | ... | 4?96 |   |----------
-------+-------------|-------------|   |   | 100?00 | 100?00 | 100?00 | +---------------------
-+------------------+-------------+-------------+
```

The results contained in these analyses are peculiarly remarkable, for they indicate the almost total absence of all those substances which the plant requires. They must, however, be considered as in a great measure exceptional cases, as it is but rarely that so large a number of constituents is absent, and it is much more frequent to find the deficiency restricted to one or two substances. They are illustrations of barrenness dependent on different circumstances. The first shows the unimportance of the organic matters of the soil, which are here unusually abundant, without in any way counteracting the infertility dependent on the absence of the other constituents. The second is that of a nearly pure sand; and the third, though it contains a greater number of the essential ingredients of the ash, is still rendered unfruitful by the deficiency of alkalies, sulphuric acid, and chlorine.

An examination of the foregoing analyses indicates pretty clearly some of the conditions of fertility of the soil, which must obviously contain all the constituents of the plants destined to grow upon it. But it by no means exhausts the subject, for numerous instances are known of soils containing all the essential elements of plants in abundance, but on which they nevertheless refuse to grow. In these instances the defect is due either to the presence of some substance injurious to the plant, or to the state of combination of those it requires being such as to prevent their absorption. Reference has been already made to the bad effects of protoxide of iron, and it would appear that organic matter is sometimes injurious. Even water, by excluding air, and so preventing those decompositions which play so important a part in liberating the essential elements from their more permanent compounds, although it cannot render a soil absolutely barren, not unfrequently materially diminishes its fertility.

The state of combination of the soil constituents unquestionably exercise a most important influence on its fertility. That this must be the case is an inference which may be easily drawn from the statements already made regarding the different minerals from which it is directly or indirectly produced. If, for instance, a soil consist to a large extent of mica, it would be found on analysis to contain abundance of potash and some other matters, and yet our knowledge of the difficulty with which that mineral is decomposed, would enable us to pronounce unfavourably of the soil; and practical experience here fully confirms the scientific inference.

The forms of combination most favourable to fertility is a subject on which our information is at present comparatively limited. It was at one time believed that solubility in water was an indispensable requisite, but recent investigations appear to lead to a directly contrary conclusion. The analyses of soils already given, show that the part directly soluble in water embraces only a certain number of the constituents of the plant, and of those dissolved the quantity is very small. This becomes still more apparent if we estimate from the analyses the actual quantities of those substances contained in an acre of soil. It is generally assumed that the soil on an imperial acre of land 10 inches deep weighs in round numbers about 1000 tons; and calculating from this, we find that the quantity of potash soluble in water in the Mid-Lothian wheat soil, amounts to no more than 70 lb. per acre. But a crop of hay carries off from the soil about 38 lb. of potash, and one of turnips, including tops, not less than 200 lb., so that if only the matters soluble in water could be taken up by the plant, such soils could not possess the amount of fertility which they are actually found to have.

It is to be remembered, also, that in these analyses the experiment is made under the most favourable circumstances for ascertaining the whole quantity of matters which are capable of dissolving in water; that practically dissolved is very different. The recent analysis by Krocker and Way of the drainage water of soils afford a means of estimating this. Way found in one gallon of the drainage water from seven different fields, collected in the end of December--

	1	2	3	4	5	6	7
Potash,	trace	trace	0?2	0?5	trace	0?2	trace
Soda,	1?0	2?7	2?6	0?7	1?2	1?0	3?0
Lime,	4?5	7?9	6?5	2?6	2?2	5?2	13?0
Magnesia,	0?8	2?2	2?8	0?1	0?1	0?3	2?0
Iron and Alumina,	0?0	0?5	0?0	none	1?0	0?5	0?0
Silica,	0?5	0?5	0?5	1?0	1?0	0?5	0?5
Chlorine,	0?0	1?0	1?7	0?1	1?6	1?1	2?2
Sulphuric acid,	1?5	5?5	4?0	1?1	1?9	3?2	9?1
Phosphoric acid,	trace	0?2	trace	trace	0?8	0?6	0?2
Ammonia,	0?18	0?18	0?18	0?12	0?18	0?06	0?18
Nitric acid,	7?7	14?4	12?2	1?5	3?5	8?5	11?5
Organic matter,	7?0	7?0	12?0	5?0	5?0	5?0	7?0

Some of the soils from which these waters were obtained had been manured with unusually large quantities of nitrogenous matters, which accounts for the large amount of nitric acid, as well as the lime which that acid has extracted. Dr. Krocker's analyses were made on soils less highly manured, and the water was collected in summer.

IN 10,000 PARTS.	1	2	3	4	5	6
Organic matter	0?5	0?4	0?6	0.06	0?3	0?6
Carbonate of lime	0?4	0?4	1?7	0?9	0?1	0?4
Sulphate of lime	2?8	2?0	1?4	0?7	0?7	0?2
Nitrate of lime	0?2	0?2	0?1	0?2	0?2	0?2
Carbonate of magnesia	0?0	0?9	0?7	0?7	0?7	0?6
Carbonate of iron	0?4	0?4	0?4	0?2	0?2	0?1
Potash	0?2	0?2	0?2	0?2	0?4	0?6
Soda	0?1	0?5	0?3	0?0	0?5	0?4
Chloride of sodium	0?8	0?8	0?7	0?3	0?1	0?1
Silica	0?7	0?7	0.06	0?5	0?6	0?5

In order to obtain from these experiments an estimate of the quantity of the substances actually dissolved, we shall select the results obtained by Way. The average rainfall in Kent, where the waters he examined were obtained, is 25 inches. Now, it appears that about two-fifths of all the rain which falls escapes through the drains, and the rest is got rid of by evaporation. An inch of rain falling on an imperial acre weighs rather more than a hundred tons; hence, in the course of a year, there must pass off by the drains about 1000 tons of drainage water, carrying with it, out of the reach of the plants, such substances as it has dissolved, and 1500 tons must remain to give to the plant all that it holds in solution. These 1500 tons of water must, if they have the same composition as that which escapes, contain only two and a half pounds of potash, and less than a pound of ammonia. It may be alleged that the water which remains, lying longer in contact with the soil, may contain a larger quantity of matters in solution; but even admitting this to be the case, it cannot for a moment be supposed that they can ever amount to more than a very small fraction of what is required for a single crop. It may therefore be stated with certainty that solubility in water is not essential to the absorption of substances by the plant, which must possess the power of itself directly attacking, acting chemically on, and dissolving them. The mode in which it does this is entirely unknown, but it in all probability depends on very feeble chemical actions, and hence the importance of having the soil constituents, not in solution, but in such a state that they may be readily made soluble by the plants. Many of the minerals from which fertile soils are formed are probably not attackable by plants when in their natural condition, and even after disintegration the quantity of the essential elements of their food, which are present in an easily assimilable state, is at no one time very large. But this is of comparatively little importance, for the soil is not an inert unchangeable substance; it is the theatre of an important series of chemical changes effected by the action of air and moisture, and producing a continued liberation of its constituents. This decomposition is effected partly by the carbonic acid of the atmosphere, but to a much larger extent by its oxygen acting upon the organic matters of the soil, and causing a constant though slow evolution of that acid, which in its turn attacks the mineral matters. Boussingault and Levy have illustrated the extent of this action by examining the composition of the air contained in the pores of different soils, and have obtained the following results:--

Nature of Soil	Crop	No. of cubic inches of air in 34 cubic inches of soil	100 VOLUMES OF AIR CONTAIN		
			Carbonic acid	Oxygen	Nitrogen
Light sandy soil, newly manured	...	8?	2?7	...	...
Do. manured 8 days before	...	...	1?4	18?0	79?6
Do. long after manuring	turnip	7?	0?3	19?0	79?7
Very sandy	Vineyard	9?	1?6	19?2	79?2
Sandy, with many stones	Forest	4?	0?7	19?1	79?2
Loamy	...	2?	0?6	...	...
Sandy, subsoil of the last	...	3?	0?4	...	...
Sandy soil, long after manuring	Trefoil	7?	0?4	19?2	80?4
Do. Recently manured	...	...	0?5	19?1	79?4
Do. manured 8 days before	...	...	1?4	18?0	79?6
Heavy clay	Jerusalem artichoke	7?	0?6	19?9	79?5
Fertile soil (moist)	Meadow	5?	1?9	19?1	78?0

From these analyses it appears that the air contained in the pores of the soil is much richer in carbonic acid than the atmosphere, the poorest soil containing about 25 times, and a recently manured soil 250 times as much. This carbonic acid, which is obviously produced by the decomposition of the vegetable matters and manure, acting partly as gas and partly dissolved in the soil water, exerts a solvent action on its constituents. And, though a very feeble acid, its continuous action produces in the course of time a large effect; while, during the interval, the constituents of the soil are safely stored up, and liberated only as the plant requires them, by which bountiful provision of nature they are exposed to fewer risks of loss than if they had been all along in a state in which they could be absorbed. Carbonic acid not only assists in effecting the decomposition of the minerals of the soil, but its aqueous solution acts as a solvent of many substances, which are quite insoluble in pure water. It is in this way that much of the lime contained in natural waters is held in solution, and it has been ascertained that magnesia, iron, and even phosphate of lime, may also be dissolved by it. It is probable that when these substances are dissolved, the plants will take them from solution in place of themselves attacking the insoluble matters; but of the extent to which this may occur nothing is yet known--the action of solvents on the soil being a subject which is as yet scarcely examined.

Carbonic acid is, however, a most important agent in producing the chemical changes in the soil, and the particular value of humus lies in its affording a supply of that substance exactly when it is wanted; but the carbonic acid of the atmosphere also takes part in these changes, although with different degrees of rapidity according to the character of the soil, acting rapidly in light, and slowly in stiff, clay soils. The solvent action of the carbonic acid is, no doubt, principally exerted on the substances soluble in acids, but not entirely, for it is known that the insoluble part is gradually being disintegrated and made soluble; and hence it is that the composition of that part of the soil which resists the action of acids, and which at first sight might appear of no moment, is really important. It is obvious that this circumstance must at once confer on the soil of the Carse of Gowrie a great superiority over those of Mid-Lothian and most other districts; for it contains in its insoluble part a quantity of alkalies which must necessarily form a source of continued fertility. Accordingly, experience has all along shown the great superiority of that soil, and of alluvial soils generally, which are all more or less similar to it. The facility with which these matters are attackable by carbonic acid is also an important element of the fertility of a soil, and it is to the existence of compounds which are readily decomposed by it that we attribute the high fertility of the trap soils.

By a further examination of the analyses of fertile soils, it is at once apparent that the most essential constituents of plants are by no means very abundant in them. In fact, phosphoric and sulphuric acids, lime, magnesia, and the alkalies, which in most instances make up nine-tenths of the ash of plants, form but a small portion of even the most fertile soils; while silica, which, except in the grasses, occurs in small quantity, oxide of iron which is a limited, and alumina a rare, constituent of the ash, constitute by far their larger part. Thus the total amount of potash, soda, lime, magnesia, phosphoric and sulphuric acids and chlorine, contained in the Mid-Lothian wheat soil amounts only to 3?888 per cent, and in the Perthshire to 6?385, the entire remainder being substances which enter into the plant for the most part in much smaller quantity. And, as these small quantities of the more important substances are capable of supplying the wants of the plant, it must be obvious that a very small fraction of the silica, oxide of iron, and alumina, which the soils contain, would afford to it the whole quantity of these substances it requires, and that the remainder must have some other functions to perform.

The soil must be considered not merely as the source of the inorganic food of plants, for it has to act also as a support for them while growing, and to retain a sufficient quantity of moisture to support their life; and unless it possess the properties which fit it for this purpose, it may contain all the elements of the food of plants, and yet be nearly or altogether barren.

The adaptation of the soil to this function is dependent to a great extent on its mechanical texture, and on this considerable light is frequently thrown by a kind of mechanical analysis.

If a soil be shaken up with water and allowed to stand for a few minutes, it rapidly deposits a quantity of grains which are at once recognised as common sand; and if the water be then poured off into another vessel and allowed to stand for a longer time, a fine soft powder, having the properties and composition of common clay, is deposited, while the clear fluid retains the soluble matters. By a more careful treatment it is possible to distinguish and separate humus, and in soils lying on chalk or limestone, calcareous matter or carbonate of lime.

In this way the components can be classified into four groups, a mixture of two or more of which in variable proportions is found in all soils.

The relative proportions in which these substances exist in soils are, as we shall afterwards see, the foundation of their classification into the light, heavy, calcareous, and other sub-divisions. But they are also intimately connected with certain chemical and mechanical peculiarities which have an important bearing on its fertility. It is a familiar fact, that particular soils are specially adapted to the growth of certain crops; and we talk of a wheat or a turnip soil as readily distinguishable. It is to be observed, however, that in many such instances the mere analysis may show no difference, or, at least, none sufficient to account for the peculiarity. A remarkable illustration is offered by the following analyses of two soils, on one of which red clover grows luxuriantly, while on the other it invariably fails.

Clover fails. Clover succeeds.

Insoluble silicates 83?0 81?4 Soluble silica 0?8 0?2 Peroxide of iron 4?5 6?8 Alumina 2?0 3?0 Lime 1?3 1?3 Magnesia 0?5 0?5 Potash 0?0 0?2 Soda 0?7 0?9 Sulphuric acid 0?5 0?8 Phosphoric acid 0?8 0?7 Carbonic acid 0?9 0?4 Chlorine trace trace Humic acid 0?2 0?3 Humine ... 0?0 Insoluble organic matters 3?0 3?1 Water 2?4 2?2 ---- ---- 99?6 100?8 Nitrogen 0?5 1?5

In this instance such difference as exists is rather in favour of the soil on which clover fails, but it is exceedingly trifling; and it is necessary to seek an explanation in the special properties of its mechanical constituents.

These properties are partly mechanical and partly chemical, and in both ways exercise an important influence on the fertility of the soil.

Sand and clay, the most important of the mechanical constituents, confer on the soil diametrically opposite properties; the former, when present in large quantity, producing what are designated as light, the latter stiff or heavy soils. The hard indestructible siliceous grains, of which sand is composed, form a soil of an open texture, through which water readily permeates; while clay, from its fine state of division, and peculiar adhesiveness or plasticity, gives it a close-textured and retentive character, and their proper intermixture produces a light fertile loam, each tempering the peculiar properties of the other. Indeed, their mixture is manifestly essential, for sand alone contains little or none of the essential ingredients of plants; and if present in large quantity, the openness of the soil is excessive, water flows through it with rapidity, manures are rapidly wasted, and on the accession of drought, the plants growing upon it soon languish and die. Clay, on the other hand, is by itself equally objectionable; the closeness of its texture prevents the spreading of the roots of plants, and the access of carbonic acid, which, as we have already seen, is so important an agent in the changes occurring in the soil. In fact, a pure clay, that is to say, a clay unmixed with sand, even though it may contain all the essential constituents of the plant, is for this reason unfertile. Practically, of course, these extreme cases rarely occur; the heaviest clay soils being mixtures of true clay with sand, and the most sandy containing their proportion of clay; but frequently the preponderance of the one over the other is so great, as to produce soils greatly inferior to those in which the mixture is more uniform.

It is easy to understand how the proportions in which sand and clay are mixed must affect the suitability of soils to particular crops, and that an open soil must be favourable to the turnip, and a heavy clay, owing to the resistance it offers to the expansion of the bulbs, unfavourable. But these substances also exercise an important chemical action on the soluble constituents of the food of plants, combining with them, and converting them into an insoluble, or nearly insoluble state, so as to prevent their being washed away by the rain or other water which percolates through the soil. It has long been known to chemists that clay has a tendency to absorb a small proportion of ammonia, and even when brought up from a great depth frequently contains that substance. It is to Mr. Thompson of Moat Hall, however, that we owe the important observation, that arable soils rapidly remove ammonia from solution, and Way, who pursued this investigation, showed that not only ammonia, but potash, and several of the other important elements of the food of plants, are thus absorbed. The removal of these substances from solution is easily illustrated by a simple experiment. It suffices to take a tall cylindrical vessel open at both ends, and filled with the soil to be operated upon, which is retained by a piece of rag tied over its lower end. A quantity of a dilute solution of ammonia being then poured upon the surface of the soil, and allowed to percolate, the first quantity which flows away is found to have entirely lost its peculiar smell and taste; and in a similar manner the removal of potash may be illustrated. This action is by no means confined to those substances when in the free state, but is equally marked when they are combined with acids in the form of salts, and in the latter case the absorption is attended with a true chemical decomposition, the base only being retained, and the acid escaping most commonly in combination with lime. Thus, if sulphate of ammonia be employed, the water which flows from the soil contains sulphate of lime, and if muriate of ammonia be used, it is muriate of lime which escapes.

This absorbent action is most remarkably manifested in the case of ammonia and potash, but it takes place also with magnesia and soda. With the latter, however, it is incomplete, only a half or a fourth of the soda being removed from solution, the difference depending to some extent on the acid with which it is in combination. The extent to which absorption takes place varies also with the nature of the soil, and the state of combination of the substance used. Exact experiments have hitherto been chiefly confined to ammonia, potash, and lime in the free state, and as bicarbonate; and the following table gives the results obtained by Way, with solutions containing about 1 per cent of these substances in solution:--

	Pure soil, Somersetshire.	Subsoil	Loamy soil, Berkshire.	Red clay.	clay,	Dorsetshire.
Ammonia, caustic	0?438	0?570	...	...		
" from muriate	0?478	0?966	0.2847	0?818		
Potash, caustic	...	...	1?50	2?87		
" from nitrate	...	...	0?980	...		
Lime, caustic	...	...	1?68	...		
" from bicarbonate	...	...	0?31	...		

From these numbers it appears that very great differences exist in the absorbent power of different soils, the first of those experimented on being capable of taking more than twice as much ammonia as the second, and nearly four times as much as the subsoil clay. It appears also, as far as absorption goes, to be immaterial whether the ammonia is free or combined. But it is different with potash, which is absorbed from the nitrate to the extent of about O? per cent, and from a caustic solution of potash to double that amount.

The circumstances under which absorption takes place modify, in a manner which cannot well be explained, the amount absorbed by the same soil. It is found generally to be most complete with very dilute solutions, and if a soil be agitated with a quantity of ammonia larger than it can take up, it will absorb only a certain amount of that substance, but by a further increase of the amount of ammonia a still larger quantity will be absorbed.

It is important to observe that when a salt is used, the base only is absorbed, and the acid escapes in combination with lime; even nitric acid, notwithstanding its importance as a food of plants, being in this predicament. From this it may be gathered that lime is not readily absorbed from solutions of its salts; indeed, it would appear that the only salt of that substance liable to absorption is the bicarbonate, from which it is taken to the extent of 1? per cent by the soil. The absorption of lime from this salt, and that of phosphoric acid, which takes place to a considerable extent, probably occurs, however, quite independently of the clay present in the soil, and is occasioned by its lime, which forms an insoluble compound with phosphoric acid, and by removing half the carbonic acid of the bicarbonate of lime converts it also into an insoluble state.

In addition to these mineral substances, organic matters are also removed from solution. This is conspicuously seen in the case of putrid urine, which not only loses its ammonia, but also its smell and colour, when allowed to percolate through soil; and an equally marked result was obtained with flax water, from which the organic matter was entirely abstracted.

The cause of this absorptive power is still very imperfectly known. Mr. Way having observed that sand has no such property, while clay, even when obtained from a considerable depth, always possesses it, supposed that the absorption was entirely due to that substance. A difficulty, however, presents itself in explaining how it should happen that while a pure clay absorbs only 0?847 of ammonia, a loamy soil, of which one-half probably is sand, should absorb a larger quantity. The inference is, that the effect cannot be due to the clay as a whole, and Mr. Way has sought to explain it by supposing that there exist in the soil particular double silicates of alumina and lime. He has shown that felspar and the other minerals from which the soil is produced have no absorbent power, but that artificial compounds can be formed which act upon solutions of ammonia and potash in a manner very similar to the soil; but there is not the slightest evidence that these compounds exist in the soil, and in the year 1853[I] I pointed out the probability that clay is not the only agent at work, but that the organic matters take part in the process. So powerful indeed is the affinity of these substances for ammonia, that chemists are at one as to the difficulty of obtaining humic and other similar acids pure, owing to the obstinacy with which they retain it; and there cannot be a doubt that in many soils these substances are in this point of view of much importance. This is particularly the case in peat soils, which, though naturally barren, may be made to produce good crops by the application of sand or gravel; and as neither of these can cause any absorption of the valuable matters, we must attribute this effect to the organic matter. Referring to an earlier series of experiments made in 1850, I showed that, if a quantity of dry peat be taken and ammonia poured on it, its smell disappears; and this may be continued until upwards of 1? per cent of dry ammonia has been absorbed, and this quantity is retained by the peat.

In this case pure ammonia was used, but Way's experiments having shown that this alkali is not absorbed from its salts by organic matters, I expressed the opinion that humate of lime (which certainly exists in most soils) ought on chemical grounds to decompose the salts of ammonia and cause the retention of their base. The recent researches of Brustlein have shown that lime does cause the organic matters to absorb ammonia from its salts. He confirms the fact that pure ammonia is absorbed by peat, and shows that decayed wood has the same effect, although both are without action on solutions of its salts. A stiff clay, on the other hand, containing organic matters and much carbonate of lime, readily absorbed ammonia, both when pure and combined; but after extracting the lime by means of a dilute acid, it lost the power of taking it from its salts, although it retained the free alkali as completely as before. On the addition of a small quantity of lime, it again acquired the power of withdrawing ammonia from its compounds. These experiments may be explained, either on the supposition of the presence of humate of lime, or by supposing that the carbonate of lime first decomposed the salts of ammonia, and that the liberated alkali combined with the organic matter. It must be admitted, however, that it is very doubtful whether the ammonia and other substances are fixed in the soil by a true chemical combination. They are certainly retained by a very feeble attraction, for it appears from Brustlein's experiments that ammonia may be, to a considerable extent, removed by washing with abundance of water, and that if the soil which has absorbed ammonia be allowed to become dry in the air, it loses half its ammonia, and after four times moistening and drying, three-fourths have disappeared. These facts are certainly not incompatible with the presence of a true chemical compound, for the humate of ammonia is not absolutely insoluble, and many cases occur of actions taking place in the presence of water, which are entirely reversed when that fluid is removed; and it is quite possible that when humate of ammonia is dried in contact with carbonate of lime, it may be decomposed, and carbonate of ammonia escape. There are other circumstances, however, which render it, on the whole, most probable that the combination is not wholly chemical, but rather of a physical character, among which may be more especially mentioned the fact, that the quantity of the substances retained by the soil is dependent on the degree of dilution of the fluid from which they are taken; and that the quantity absorbed never exceeds a very small fraction of the weight of the soil.

The practical inferences to be drawn from these facts regarding the value of soils are of the highest importance. It is obvious that two soils having exactly the same chemical composition may differ widely in absorptive power, and that which possesses it most largely must have the highest agricultural value. The examination of different soils, in this point of view, is a subject of much importance, and deserves the best attention of both farmers and chemists, although little has as yet been done in regard to it, and the results which have been obtained are not of a very satisfactory character. Liebig states, that in his experiments, all the arable soils examined possessed the same absorptive power, whether they contained a large or a small proportion of lime or alumina. It can scarcely be expected, however, that this should be true in all cases, and there are many facts which seem to indicate that differences must exist. It is well known that there are some soils in which the manure is very rapidly exhausted, and it is more than probable that this effect is due to deficient absorptive power, which leaves the soluble matters at the mercy of the weather, and liable at any moment to be washed out by a heavy fall of rain.

The more strictly mechanical properties of the soil, such as its relations to heat and moisture, are not less important than its chemical composition. It is known that soils differ so greatly in these respects as sometimes materially to affect their productive capacity. Thus, for instance, two soils may be identical in composition, but one may be highly hygrometric, that is, may absorb moisture readily from the air, while the other may be very deficient in that property. Under ordinary circumstances no difference will be apparent in their produce, but in a dry season the crop upon the former may be in a flourishing condition, while that on the latter is languishing and enfeebled, merely from its inability to absorb from the air, and supply to the plant the quantity of water required for its growth. In the same way, a soil which absorbs much heat from the sun's rays surpasses another which has not that property; and though in many cases this effect is comparatively unimportant, in others it may make the difference between successful and unsuccessful cultivation in soils which lie in an unfavourable climate or exposure.

The investigation of the physical characters of soils has attracted little attention, and we owe all our present knowledge of the subject to a very elaborate series of researches on this subject, published by Sch黙甘ler, nearly thirty years ago. He determined 1st, The specific gravity of the soils; 2d, The quantity of water which they are capable of imbibing; 3d, The rapidity with which they give off by evaporation the water they have imbibed; that is, their tendency to become dry; 4th, The extent to which they shrink in drying; 5th, Their hygrometric power; 6th, The extent to which they are heated by the sun's rays; 7th, The rapidity with which a heated soil cools down, which indicates its power of retaining heat; 8th, Their tenacity, or the resistance they offer to the passage of agricultural implements; 9th, Their power of absorbing oxygen from the air. Each of these experiments was performed on several different soils, and on their mechanical constituents. Sch黙甘ler's experiments are undoubtedly important, and though the methods employed are some of them not altogether beyond cavil, they have apparently been performed with great care. It is nevertheless desirable that they should be repeated, for such facts ought not to rest on the authority of one experimenter, however skilful and conscientious, nor on a single series of soils, which may not give a fair representation of their general physical properties. In fact, Sch黙甘ler appears to imagine that having once determined the extent to which the sand, clay, and other mechanical constituents of the soil possess these properties, we are in a condition to predicate the effect of their mixture in variable proportions, although this is by no means probable.

In examining these properties, Sch黙甘ler selected for experiment, pure siliceous sand, calcareous sand (carbonate of lime in coarse grains), finely powdered carbonate of lime, pure clay, humus, and powdered gypsum. He used also a heavy clay consisting of 11 per cent of sand and 89 of pure clay, a somewhat stiff clay containing 24 per cent of sand and 76 of clay, a light clay with 40 per cent of sand and 60 of pure clay, a garden soil consisting of 52?

per cent of clay, 36? of siliceous sand, 1? of calcareous sand, 2 per cent of finely divided carbonate of lime, and 7? of humus, and two arable soils, one from Hoffwyl, and one from a valley in the Jura, the former a somewhat stiff, the latter a light soil.

soil	Specific gravity.	Water absorbed by 100 parts at 66?	Diminution in bulk per cent.	parts of water evaporate during drying in four hours	Of 100 parts absorbed moist soil
Siliceous sand	2?53	25	88?	0?	
Light clay	2?01	40	52?	6?	
Heavy clay	2?03	61	34?	11?	
Carbonate of lime	2?68	85	28?	5?	
Gypsum	2?58	27	71?	0?	
Soil from Hoffwyl	2?01	52	32?	12?	
Calcareous sand	2?22	29	75?	0?	
Stiff clay	2?52	50	45?	8?	
Pure clay	2?91	70	31?	18?	
Humus	1?25	190	20?	20?	
Garden soil	2?32	96	24?	14?	
Soil from Jura	2?26	47	40?	9?	

Power of retaining hygrometric water absorbed by 77?65 grains of the soil spread on a surface of 141?8 square inches.

	12 hours.	24 hours.	48 hours.	72 hours.	Quantity of heat.	Tenacity Pure clay, 100.
Siliceous sand	0	0	0	0		95?
Calcareous sand	0?54	0?31	0?31	0?31		100?
Light clay	1?17	2?02	2?56	2?56		76?
Stiff clay	1?25	2?10	2?18	2?95		71?
Heavy clay	2?10	2?72	3?80	3?57		68?
Pure clay	2?49	3?34	3?96	3?73		66?
Carbonate of lime	2?02	2?87	2?95	2?95		61?
Humus	6?60	7?69	8?70	9?40		49?
Gypsum	0?77	0?77	0?77	0?77		73?
Garden soil	2?95	3?65	3?50	4?04		64?
Soil from Hoffwyl	1?32	1?71	1?71	1?71		70?
Soil from Jura	1?78	1?63	1?40	1?40		74?

Quantity of oxygen absorbed by 77?65 grains of the moist soil in 30 days, from 15 cubic inches of the atmospheric air. Expressed in cubic inches.

				Siliceous sand	0	0?4
Calcareous sand	0	0?4		Light clay	57?	1?9		Stiff clay	68?	1?5
Heavy clay	83?	2?4		Pure clay	100?	2?9		Carbonate of lime	5?	1?2		Humus	8?	3?4		Gypsum	7?	0?0		Garden soil	7?	2?0
Soil from Hoffwyl	33?	2?3		Soil from Jura	22?	2?5

The experiments detailed in the preceding table speak in a great measure for themselves, and scarcely require detailed comment. It may be remarked, however, that the columns illustrating the relations of the soil to water are probably more important than the others. The superiority of a retentive over an open soil is sufficiently familiar in practice, and though this is no doubt partly due to the former absorbing and retaining more completely the ammonia and other valuable constituents of the manures applied to it, it is also dependent to an equal if not greater extent upon the power it possesses of retaining moisture. A reference to the table makes it apparent that this power is presented under three different heads, which are certainly related to one another, but are not identical. In the second column of the table is given the quantity of water absorbed by the soil, determined by placing a given weight of the perfectly dry soil in a funnel, the neck of which is partially stopped with a small piece of sponge or wool, pouring water upon it, and weighing it after the water has ceased to drop from it. This may be considered as representing the quantity of water retained by these different soils when thoroughly saturated by long continued rains. The column immediately succeeding gives the quantity of that water which escapes by evaporation from the same soil after exposure for four hours to dry air at the temperature of 66? The fifth, sixth, seventh, and eighth columns indicate the quantity of moisture absorbed, when the soil, previously artificially dried, is exposed to moist air for different periods. These characters are dependent principally, though not entirely, on the porosity of the soil. The last may also be in some measure due to the presence of particular salts, such as common salt, which has a great affinity for moisture, but is chiefly occasioned by their peculiar structure. It is to be remarked that clay and humus are two of the most highly hygrometric substances known, and it is peculiarly interesting to observe, that by a beneficent provision of nature, they also form a principal part of all fertile soils. The quantity of water imbibed by the soil is important to its fertility, in so far as it prevents it becoming rapidly dry after having been moistened by the rains. It is valuable also in another point of view, because if the soil be incapable of absorbing much water, it becomes saturated by a moderate fall of rain, and when a larger quantity falls, the excess of necessity percolates through the soil, and carries off with it a certain quantity of the soluble salts. Important as this property is, however, it must not be possessed in too high a degree, but must permit the evaporation of the water retained with a certain degree of rapidity. Soils which do not admit of this taking place are the cause of much inconvenience and injury in practice. By becoming thoroughly saturated with moisture during winter, they remain for a long time in a wet and unworkable condition, in consequence of which they cannot be prepared and sown until late in the season, and though chemically unexceptionable, they are always disadvantageous, and in some seasons greatly disappoint the hopes of the farmer.

The extent to which the imbibition and evaporation of water takes place is very variable, but they are obviously related to one another, the soils which absorb it least abundantly parting with it again with the greatest, facility; for it appears that siliceous sand absorbs only one-fourth of its weight of water, and again gives off in the course of four hours four-fifths of that it had taken up, while humus, which imbibes nearly twice its weight, retains nine-tenths of that quantity after four hours' exposure. Long-continued and slow evaporation of the water absorbed by a soil is injurious in another way, for it makes the soil "cold"--a term of practical origin, but which very correctly expresses the peculiarity in question. It is due to the fact, that when water evaporates it absorbs a very large quantity of heat, which prevents the soil acquiring a sufficiently high temperature from the sun's rays. The soils which have absorbed a large quantity of moisture shrink more or less in the process of drying, and form cracks, which often break the delicate fibres of the roots of the plants, and cause considerable injury: the extent of this shrinking is given in the fourth column.

The relation of the soils to heat divides itself into two considerations: the amount of heat absorbed by the soil, and the degree in which it is retained. Of these the latter only is illustrated in the table. The former is dependent on so many special considerations, that the results cannot be tabulated in a satisfactory manner. It is independent of the chemical nature of the soil, but varies to a great extent according to its colour, the angle of incidence of the sun's rays, and its state of moisture. It is, however, an important character, and has been found by Girardin to exercise a considerable influence on the rapidity with which the crop ripens. He found in a particular year that, on the 25th of August, 26 varieties of potatoes were ripe on a very dark-coloured sandy vegetable mould, 20 on an ordinary sandy soil, 19 on a loamy soil, and only 16 on a nearly white calcareous soil.

The tenacity of the soil is very variable, and indicates the great differences in the amount of power which must be expended in working them. According to Sch 尉 ler, a soil whose tenacity does not exceed 10, is easily tilled, but when it reaches 40 it becomes very difficult and heavy to work.

On examining the table it becomes manifest, that as far as its mechanical properties are concerned, humus is a substance of the very highest importance, for it confers on the soil, in a high degree, the power of absorbing and retaining water, diminishes its tenacity and permits its being more easily worked, adds to its hygrometric power and property of absorbing oxygen from the air, and finally, from its dark colour, causes the more rapid absorption of heat from the sun's rays. It will be thus understood, that though it does not directly supply food to the plant, it ministers indirectly in a most important manner to its well-being, and that to so great an extent that it must be considered an indispensable constituent of a fertile soil. But it is important to observe that it must not be present in too large a quantity, for an excess does away with all the good effects of a smaller supply, and produces soils notorious for their infertility.

Such are the important physical properties of the soil, and it is greatly to be desired that they should be more extensively examined. The great labour which this involves has, however, hitherto prevented its being done, and will, in all probability, render it impossible except in a limited number of cases. Some of these characters are, however, of minor importance, and for ordinary purposes it might be sufficient to determine the specific gravity of the soil in the dry and moist state, the power of imbibing and retaining water, its hygrometric power, its tenacity, and its colour. With these data we should be in a condition to draw probable conclusions regarding the others; for the higher the specific gravity in the dry state, the greater is the power of the soil to retain heat, and the darker its colour the more readily does it absorb it. The greater its tenacity the more difficult is it to work, and the greater difficulty will the roots of the young plant find in pushing their way through it. The greater the power of imbibing water, the more it shrinks in drying; and the more slowly the water evaporates, the colder is the soil produced. The hygrometric power is so important a character that Davy and other chemists have even believed it possible to make it the measure of the fertility of a soil; but though this may be true within certain limits, it must not be too broadly assumed, the results of recent experiments by no means confirming the opinion in its integrity, but indicating only some relation between the two.

The Subsoil.--The term soil is strictly confined to that portion of the surface turned over by the plough working at ordinary depth; which, as a general rule, may be taken at 10 inches. The portion immediately subjacent is called the subsoil, and it has considerable agricultural importance, and requires a short notice. In many instances, soil and subsoil are separated by a purely imaginary line, and no striking difference can be observed either in their chemical or physical characters. In such cases it has been the practice with some persons not to limit the term soil to the upper portion, but to apply it to the whole depth, however great it may be, which agrees in characters with the upper part, and only to call that subsoil which manifestly differs from it. This principle is perhaps theoretically the more correct, but great practical advantages are derived from limiting the name of soil to the depth actually worked in common agricultural operations. The subsoil is always analogous in its general characters to a soil, but it may be either identical with that which overlies it or not. Of the former, striking illustrations are seen in the wheat subsoils, the analyses of which have been already given. In the latter case great differences may exist, and a heavy clay is often found lying on an open and porous sand, or on peat, and vice versa. Even where the characters of the subsoil appear the same as those of the soil, appreciable chemical differences are generally observed, especially in the quantity of organic matter, which is increased in the soil by the decay of plants growing upon it and by the manure added. In general, then, all that we have said regarding the characters of soils both chemically and physically, will apply to the subsoils, except that, owing to the difficulty with which the air reaches the latter, some minor peculiarities are observed. The most important is the effect of the decay of vegetable matter, without access of air, which is attended by the reduction of the peroxide of iron to the state of protoxide, and not unfrequently by the production of sulphuret of iron, compounds which are extremely prejudicial to vegetation, and occasionally give rise to some difficulties when the subsoil is brought to the surface, as we shall afterwards have to notice.

The physical characters of the subsoil are often of much importance to the soil itself. As, for instance, where a light soil lies on a clay subsoil, in which case its value is much higher than if it reposed on an open or sandy subsoil. And in many similar modes an important influence is exerted; but these belong more strictly to the practical department of agriculture, and need not be mentioned here.

Classification of Soils.--Numerous attempts have been made to form a classification of soils according to their characters and value, but they have not hitherto proved very successful; and the result of more recent chemical investigations has not been such as to encourage a farther attempt. We have not at present data sufficient for the purpose, nor, if we had, would it be possible to arrange any soil in its class except after an elaborate chemical examination. The only classification at present possible must be founded on the general physical characters of the soil; and the ordinary mode followed in practice of dividing them into clays, loams, etc. etc., which we need not here particularize, fulfils all that can be done until we have more minute information regarding a large number of soils. Those of our readers who desire more full information on this point are referred to the works of Thaer, Sch 黜 ler, and others, where the subject is minutely discussed.

FOOTNOTES:

[Footnote I: Transactions of the Highland and Agricultural Society, vol. vi., p. 317.]

CHAPTER VI.

THE IMPROVEMENT OF THE SOIL BY MECHANICAL PROCESSES.

Comparatively few uncultivated soils possess the physical properties or chemical composition required for the production of the most abundant crops. Either one or more of the substances essential to the growth of plants are absent, or, if present, they are deficient in quantity, or exist in some state in which they cannot be absorbed. Such defects, whether mechanical or chemical, admit of diminution, or even entire removal, by certain methods of treatment, the adaptation of which to particular cases is necessarily one of the most important branches of agricultural practice, as the elucidation of their mode of action is of its theory. The observations already made with regard to the characters of fertile soils must have prepared the reader for the statement that these defects may be removed, either by mechanical or chemical processes. The former method of improvement may at first sight appear to fall more strictly under the head of practical agriculture, of which the mechanical treatment of the soil forms so important a part, and that their improvement by chemical means should form the sole subject of our consideration in a treatise on agricultural chemistry. But the line of demarcation between the mechanical and the chemical, which seems so marked, disappears on more minute observation, and we find that the mechanical methods of improvement are frequently dependent on chemical principles; and those which, at first sight, appear to be entirely chemical, are also in reality partly mechanical. It will be necessary for us, therefore, to consider shortly the mechanical methods of improving the soil.

Draining.--By far the most important method of mechanically improving the soil is by draining--a practice the beneficial action of which is dependent on a great variety of circumstances. It is unnecessary to insist on the advantage derived from the rapid removal of moisture, which enables the soil to be worked at times when this used to be almost impossible, and other direct practical benefits. Of its more strictly chemical effects, the most important is probably that which it produces on the temperature of the soil. It has been already remarked that the germination of a seed is dependent on the soil in which it is sown acquiring a certain temperature, and the rapidity of the after-growth of the plant is, in part at least, dependent on the same circumstance. The necessary temperature is speedily attained by the heating action of the sun's rays, when the soil is dry; but when it is wet, the heat is expended in evaporating the moisture with which it is saturated; and it is only after this has been effected that it acquires a sufficiently high temperature to produce the rapid growth of the seeds committed to it.

The extent to which this effect occurs may be best illustrated by reference to some experiments made by Sch 黜 ler, in which he determined the temperature attained by different soils, in the wet and dry state, when exposed to the sun's rays, from 11 till 3 o'clock, in the latter part of August, when the temperature in the shade varied from 73?to 77?

Description of Soil	Wet	Dry
	Degs.	Degs.
Siliceous sand	99?	112?
Calcareous sand	99?	112?
Sandy clay	98?	111?
Loamy clay	99?	112?
Stiff clay	99?	112?
Fine bluish-grey clay	99?	113?
Garden mould	99?	113?
Arable soil	97?	111?
Slaty marl	101?	115?

In a soil which is naturally dry or has been drained, the superfluous moisture escapes by the drains, and only that comparatively small quantity which is retained by capillary attraction is evaporated, and hence the soil is more frequently and for a longer period in a condition to take advantage of the heating effect of the sun's rays, and in this way the period of germination, and, by consequence also, that of ripening is advanced. The extent of this influence is necessarily variable, but it is generally considerable, and in some districts of Scotland the extensive introduction of draining has made the harvest, on the average of years, from ten to fourteen days earlier than it was before. It is unnecessary to insist on the importance of such a change, which in upland districts may make cultivation successful when it was previously almost impossible. The removal of moisture by drainage affects the physical characters of the soil in another manner; it makes it lighter, more friable, and more easily worked; and this change is occasioned by the downward flow of the water carrying with it to the lower part of the soil the finer argillaceous particles, leaving the coarser and sandy matters above, and in this way a marked improvement is produced on heavy and retentive clays. The access of air to the soil is also greatly promoted by draining. In wet soils the pores are filled with water, and hence the air, which is so important an agent in their amelioration, is excluded; but so soon as this is removed, the air is enabled to reach and act upon the organic matters and other decomposable constituents present. In this way also provision is made for the frequent change of the air which permeates the soil; for every shower that falls expels from it a quantity of that which it contains, and as the moisture flows off by the drains, a new supply enters to take its place, and thus the important changes which the atmospheric oxygen produces on the soil are promoted in a high degree. The air which thus enters acts on the organic matters of the soil, producing carbonic acid, which we have already seen is so intimately connected with many of its chemical changes. In its absence the organic matters undergo different decompositions, and pass into states in which they are slowly acted on, and are incapable of supplying a sufficient quantity of carbonic acid to the soil; and they thus exercise an action on the peroxide of iron, contained in all soils, reduce it to the state of protoxide, or, with the simultaneous reduction of the sulphuric acid, they produce sulphuret of iron, forms of combination which are well known to be most injurious to vegetation.

The removal of water from the lower part of the soil, and the admission of air, which is the consequence of draining, submits that part of it to the same changes which take place in its upper portion, and has the effect of practically deepening the soil to the extent to which it is thus laid dry. The roots of the plants growing on the soil, which stop as soon as they reach the moist part, now descend to a lower level, and derive from that part of it supplies of nourishment formerly unavailable. The deepening of the soil has further the effect of making the plants which grow upon it less liable to be burned up in seasons of drought, a somewhat unexpected result of making a soil drier, but which manifestly depends on its permitting the roots to penetrate to a greater depth, and so to get beyond the surface portion, which is rapidly dried up, and to which they were formerly confined.

It may be added also that the abundant escape of water from the drains acts chemically by removing any noxious matters the soil may contain, and by diminishing the amount of soluble saline matters, which sometimes produce injurious effects. It thus prevents the saline incrustation frequently seen in dry seasons on soils which are naturally wet, and which is produced by the water rising to the surface by capillary attraction, and, as it evaporates, depositing the soluble substances it contained, as a hard crust which prevents the access of air to the interior of the soil.

It is thus obvious that the drainage of the soil modifies its properties both mechanically and chemically. It exerts also various other actions in particular cases which we cannot here stop to particularize. It ameliorates the climate of districts in which it is extensively carried out, and even affects the health of the population in a favourable manner. The sum of its effects must necessarily differ greatly in different soils, and in different districts; but a competent authority[J] has estimated, that, on the average, land which has been drained produces a quarter more grain per acre than that which is undrained. But this by no means exhausts the benefits derived from it, draining being merely the precursor of further improvement. It is only after it has been carried out that the farmer derives the full benefit of the manures which he applies. He gains also by the increased facility of working the soil, and by the rapidity with which it dries after continued rain, thus enabling him to proceed at their proper season with agricultural operations, which would otherwise have to be postponed for a considerable time.

It would be out of place to enlarge here upon the mode in which draining ought to be carried out; it may be remarked, however, that much inconvenience and loss has occasionally been produced by too close adherence to particular systems. No rules can be laid down as to the depth or distance between the drains which can be universally applicable, but the intelligent drainer will seek to modify his practice according to the circumstances of the case. As a general rule, the drains ought to be as deep as possible, but in numerous instances it may be more advantageous to curtail their depth and increase their number. If, for instance, a thick impervious pan resting on a clay were found at the depth of three feet below the surface, it would serve no good purpose to make the drains deeper; but if the pan were thin, and the subjacent layer readily permeable by water, it might be advantageous to go down to the depth of four feet, trusting to the possible action of the air which would thus be admitted, gradually to disintegrate the pan, and increase the depth of soil above it. It is a common opinion that if we reach, at a moderate depth, a tenacious and little permeable clay, no advantage is obtained by sinking the drains into it; but this is an opinion which should be adopted with caution, both because no clay is absolutely impermeable, even the most tenacious permitting to a certain extent the passage of water, and because the clay may have been brought down by water from the upper part of the soil, and may have stopped there merely for want of some deeper escape for the water, and which drains at a lower level might supply. In some cases it may even be advisable to vary the depth of the drains in different parts of the same field, and the judicious drainer may sometimes save a considerable sum by a careful observation of the peculiarities of the different parts of the ground to be drained.

Subsoil and Deep Ploughing.--It frequently happens, when a soil is drained, that the subsoil is so stiff as to permit the passage of water imperfectly, and to prevent the tender roots of the plant from penetrating it, and reaching the new supplies of nourishment which are laid open to them. In such cases the benefits of subsoil ploughing and deep ploughing are conspicuous. The mode of action of these two methods of treatment is similar but not identical. The subsoil plough merely stirs and opens the subsoil, and permits the more ready passage of water and the access of air and of the roots of plants--the former to effect the necessary decompositions, the latter to avail themselves of the valuable matters set free. But deep ploughing produces more extensive changes; it raises new soil to the surface, mixes it with the original soil, and thus not only brings up fresh supplies of valuable matters to it, but frequently changes its chemical and mechanical characters, rendering a heavy soil lighter by the admixture of a light subsoil, and vice versa. Both are operations which are useless unless they are combined with draining, for it must manifestly serve no good purpose to attempt to open up a soil unless the water which lies in it be previously removed. In fact, subsoiling is useless unless the subsoil has been made thoroughly dry; and it has been found by experience that no good effects are obtained if it be attempted immediately after draining, but that a sufficient time must elapse, in order to permit the escape of the accumulated moisture, which often takes place very slowly. Without this precaution, the subsoil, after being opened by the plough, soon sinks together, and the good effects anticipated are not realized. The necessity for allowing some time to elapse between draining and further operations is still more apparent in deep ploughing, when the soil is actually brought to the surface. In that case it requires to be left for a longer period after draining, in order that the air may produce the necessary changes on the subsoil; for if it be brought up after having been for a long time saturated with moisture, and containing its iron as protoxide, and the organic matter in a state in which it is not readily acted upon by the air, the immediate effect of the operation is frequently injurious in place of being advantageous. One of the best methods of treating a soil in this way is to make the operation a gradual one, and by deepening an inch or two every year gradually to mix the soil and subsoil; as in this way from a small quantity being brought up at a time no injurious effects are produced. Deep ploughing may be said to act in two ways, firstly, by again bringing to the surface the manures which have a tendency to sink to the lower part of the soil, and, secondly, by bringing up a soil which has not been exhausted by previous cropping--in fact a virgin soil.

The success which attends the operation of subsoiling or deep ploughing must manifestly be greatly dependent on the character of the subsoil, and good effects can only be obtained when its chemical composition is such as to supply in increased quantity the essential constituents of the plant; and it is no doubt owing to this that the opinions entertained by practical men, each of whom speaks from the results of his own experience, are so varied. The effects produced by deep ploughing on the estates of the Marquis of Tweeddale, are familiarly known to most Scottish agriculturists, and they are at once explained by the analyses of the soil and subsoil here given, which show that the latter, though poor in some important constituents, contains more than twice as much potash as the soil.

Soil. Subsoil.

Insoluble silicates 87·23 82·2 Soluble silica 0·93 0·2 Alumina and peroxide of iron 4·29 8·0 Lime 0·41 0·8 Magnesia 0·90 0·4 Sulphuric acid 0·27 0·3 Phosphoric acid 0·40 trace Potash 0·52 0·2 Soda 0·50 0·4 Water 1·56 3·6 Organic matter 5·20 4·2 ------ ----- 100·21 99·3

In addition to the difference in the amount of potash, something is probably due to the large proportion of alumina and oxide of iron in the subsoil, which for this reason must be more tenacious than the soil itself, which appears to be rather light. In other instances, the use of the subsoil plough has occasioned much disappointment, and has led to its being decried by many practical men; but of late years its use having become better understood, its merits are more generally admitted. We believe, that in all cases in which the soil is deep, more or less marked good effects must be produced by its use, but of course there must be cases in which, from the defective composition of the subsoil or other causes, it must fail. It may sometimes be possible a priori to detect these cases, but in a large majority of them our knowledge is still too limited to admit of satisfactory conclusions being arrived at.

Improving the Soil by Paring and Burning.--It has long been familiarly known, that a decided improvement has been produced on some soils by burning. Its advantages have chiefly been observed on two sorts, heavy clays and peat soils, on both of which it has been practised to a great extent. The action of heat on the heavy clays appears to be of a twofold character, depending partly on the change effected in its physical properties, and partly on a chemical decomposition produced by the heat. The operation of burning is effected by mixing the clay with brushwood and vegetable refuse, and allowing it to smoulder in small heaps for some time. It is a process of some nicety, and its success is greatly dependent on the care which has been taken to keep the temperature as low as possible during the whole course of the burning.

Experience has shown that burning is by no means equally advantageous to all clays, but is most beneficial on those containing a considerable quantity of calcareous matter, and of silicates of potash. In such clays heat operates by causing the lime to decompose the alkaline silicates, and liberate a quantity of the potash which was previously in an unavailable state. Its effect may be best illustrated by the following analyses by Dr. Voelcker of a soil, and the red ash produced in burning it.

Soil. Red Ash. Water 0·3 1·8 Organic matter 10·7 3·2 Oxides of iron and alumina 13·0 18·2 Carbonate of lime 23·0 8·3 Sulphate of lime trace 1·5 Carbonate of magnesia 1·0 " Magnesia " 1·6 Phosphoric acid trace 0·1 Potash 0·8 1·8 Soda 0·3 " Chloride of sodium " 1·3 Insoluble matter, chiefly clay 49·6 62·2 ----- ----- 100·7 100·0

In this instance the quantity of burned soil amounted to about fifteen tons per acre, and it is obvious that the quantity of potash which had been liberated from the insoluble clay and the phosphoric acid are equal to that contained in a considerable manuring. In order to obtain these results, it is necessary, as has been already observed, to keep the temperature as low as possible during the process of burning, direct experiment having shown that when this precaution is not observed another change occurs, whereby the potash, which at low temperatures becomes soluble, passes again into an insoluble state. A part of the beneficial effect is no doubt also due to the change produced in the physical characters of the clay by burning, which makes it lighter and more friable, and by mixture with the unburnt clay ameliorates the whole. This improvement in the physical characters of the clay also requires that it shall be burnt with as low a heat as possible; for if it rises too high, the clay coheres into hard masses which cannot again be reduced to powder, and the success of the operation of burning may always be judged of by the readiness with which it falls into a uniform friable powder.

The improvement of peat by burning has been practised to some extent in Scotland, though less frequently of late years than formerly; but it is still the principal method of reclaiming peat soils in many countries, and particularly in Finland, where large breadths of land have been brought into profitable cultivation by means of it. The modus operandi of burning peat is very simple; it acts by diminishing the superabundant quantity of humus or other organic matters, which, in the previous section we have seen to be so injurious to the fertility of the soil. It may act also in the same way as it does on clay, by making part of the inorganic constituents more really soluble, although it is not probable that its effect in this way can be very marked. Its chief action is certainly by destroying the organic matters, and by thus improving the physical character of the peat, and causing it to absorb and retain a smaller quantity of water than it naturally does. For this reason it is that it proves successful only on thin peat bogs, for if they be deep, the inorganic matters soon sink into the lower part, and the surface relapses into its old state of infertility. It is probably for this reason that the practice has been so much abandoned in Scotland, more especially as other and more economical modes of treating peat soils have come into use.

Warping.--This name has been given to a method of improving soils by causing the water of rivers to deposit the mud it carries in suspension upon them, and which has been largely practised in the low lying lands of Lincoln and Yorkshire, where it was introduced about a century ago. It is most beneficial on sandy or peaty soil, and by its means large tracts of worthless land have been brought under profitable cultivation. It requires that the land to be so treated shall be under the level of the river at full tide, and it is managed by providing a sluice through which the river water is allowed to flood the land at high tide, and again to escape at ebb, leaving a layer of mud generally about a tenth of an inch in thickness, which it brought along with it. By the repetition of this process, a layer of several feet in thickness, of an excellent soil, is accumulated on the surface. Herapath, who has carefully examined this subject chemically, has shown that in one experiment where the water used contained 233 grains of mud per gallon, 210 were deposited during the warping. The following analyses will show the general nature of the matters deposited, and the change which they produce on the soil:--No. 1 is the mud from the Humber in its natural state, No. 2 a specimen of warp of average quality artificially dried, No. 3 a sandy soil before warping, and No. 4, the same fifteen years after having received a coating of 11 inches of mud.

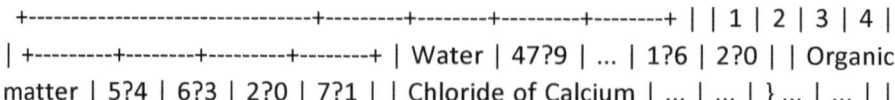

	1	2	3	4
Water	47?9	...	1?6	2?0
Organic matter	5?4	6?3	2?0	7?1
Chloride of Calcium	...	...	} ...	...

Magnesium	} ...	0?0	} ...	...		
Sodium	} 1?6	} 0?4	} 0?4	0?6		
Potassium	}	}	}			
Sulphate of Soda	} ...	0?1	} ...	...		
Magnesia	} ...	1?8	} ...	...		
Lime	trace	1?0	trace	trace		
Carbonate of Magnesia	2?0	0?1	trace	0?9		
Lime	3?9	8?8	trace	0?6		
Potash and Soda	0?8	0?7	trace	0?7		
Magnesia	1?9	2?0	trace	0?6		
Lime	0?9	0?8	trace	0?4		
Peroxide of Iron	} 6?3	5?5	0?8	1?7		
Alumina	}	8?8	0?9	0?1		
Phosphate of Iron		0?8	1?4	trace	0?8	
Silica	...	9?5	0?4	2?7		
Sand and Stones	29?5	55?7	95?1	84?7		
	100?0	100?0	100?0	100?0		

It is easy to understand the importance of the effects produced by adding to any soil large quantities of a mud containing upwards of one per cent of phosphate of iron; and in point of fact, Herapath has calculated that in one particular instance the quantity of phosphoric acid brought by warping upon an acre of land, exceeded seven tons per acre. As, moreover, the matters are all in a high state of division, they must exist in a condition peculiarly favourable to the plant. The overflow of the Nile is only an instance of warping on the large scale, with this difference, that it is repeated once only in every year, whereas, in this country, the operation is repeated at every tide until a deposit sometimes of several feet in thickness is obtained, after which it is stopped, and the soil brought under ordinary cultivation.

An operation which is, in some respects, the converse of warping, has been carried out on Blair-Drummond Moss, where the peat has been dislodged and carried off by the action of water, leaving the subjacent soil in a state fitted for cropping. Of course both this and warping are restricted to special localities, but they are most important means of ameliorating the soil when circumstances admit of their being carried out.

Mixing of Soils.--When soils possess conspicuous defects in their physical, and even in their chemical properties, great advantages may, in some instances, be derived from their proper admixture. A light sandy soil, for instance, is greatly improved by the addition of clay, and vice versa; so that, when two soils of opposite properties occur near to one another, both may be improved by mixture. It has been applied to the improvement of heavy clay soil and of peat, the former being mixed with sand or marl so as to diminish its tenacity; the latter with clay or gravel to add to its inorganic matters, and in both instances it has proved successful.

The process of chalking, which has been carried out on a large scale in some parts of England, and which consists in bringing up the chalk from pits, penetrating through the overlying tenacious clay, and mixing it with the soil, operates, to some extent, in a similar manner, though no doubt the lime also exercises a strictly chemical action. It is probable that the mixing of soils might be advantageously extended, and it merits more minute study than it has yet obtained. Its use is obviously limited by the expense, because, of course, where good effects are to be obtained, it is necessary to remove large quantities of soil, in some instances as much as 50 or 100 tons per acre, but the expense might be much diminished if it were carried out methodically, and on a considerable scale. The admixture of highly fertile soils with others of inferior quality is also worthy of attention; indeed, it is understood that this has been done, to some extent, with the rich trap soils of some parts of Scotland, but the extent of the benefit derived from it has not been made public.

FOOTNOTES:

[Footnote J: Mr. Dudgeon, Spylaw. Transactions of the Highland Society, vol. cxxix., p. 505.]

CHAPTER VII.

THE GENERAL PRINCIPLES OF MANURING.

In their natural condition all soils not absolutely barren are capable of supporting a certain amount of vegetation, and they continue to do so for an unlimited period, because the whole of the substances extracted from them are again restored, either directly by the decay of the plants, or indirectly by the droppings of the wild animals which have browzed upon them. Under these circumstances, a soil yields what may be called its normal produce, which varies within comparatively narrow limits, according to the nature of the season, temperature, and other climatic conditions. But the case is completely altered if the crop, in place of being allowed to decay on the soil, is removed from it, for, though the air will continue to afford an undiminished supply of those elements of the food of plants which may be derived from it, the fixed substances, which can only be obtained from the soil, decrease in quantity, and are at length entirely exhausted. In this way a gradual diminution of the fertility of the soil takes place, until, after the lapse of a period, longer or shorter, according to its natural resources, it will become entirely incapable of maintaining a crop, and fall into absolute infertility unless the substances removed from it are restored from some other source in the form of manure. When this is done, the fertility of the soil may not only be sustained but greatly increased, and, in point of fact, all cultivated soils, by the use of manure, are made to yield a much larger crop than they can do in their natural condition.

The fundamental principle upon which a manure is employed is that of adding to the soil an abundant supply of the elements removed from it by plants in the condition best fitted for absorption by their roots; but looked at in its broadest point of view, it acts not merely in this way, but also by promoting the decomposition of the already partially disintegrated rocks of which the soil is composed, setting free those substances it already contains, and facilitating their absorption by the plants.

In considering the practical applications of the broad general principle just stated, it might be assumed that a manure ought invariably to contain all the elements of plants in the quantities in which they are removed by the crops, and that when this has been accurately ascertained by analysis, it would only be necessary to use the various substances in the proportions thus indicated. But this, though a very important, and no doubt in many cases essential condition, is by no means the only matter which requires to be taken into consideration in the economical application of manures. And this becomes sufficiently obvious when the circumstances attending the exhaustion of the soil are minutely examined. When a soil is cropped during a succession of years with the same plant, and at length becomes incapable of longer maintaining it, the exhaustion is rarely, if ever, due to the simultaneous consumption of all its different constituents, but generally depends upon that of one individual substance, which, from its having originally existed in the soil in comparatively small quantity, is removed in a shorter time than the others. To restore the fertility of a soil in this condition, it is by no means necessary to supply all the different substances required by the plant, for it will suffice to add that which has been entirely removed. On the other hand, if an ordinary soil be supplied with a manure containing a very small quantity of one of the elements of plant food, along with abundance of all the others, the amount of increase which it yields must obviously be measured, not by those which are abundant, but by that which is deficient; for the crop which grows luxuriantly so long as it obtains a supply of all its constituents, is arrested as effectually by the want of one as of all, as has been proved by the experiments of Prince Salm Horstmar and others, referred to in a previous chapter; and hence, in order to obtain a good crop, it would be necessary to use the manure in such abundance as to supply a sufficiency of the deficient element for that purpose. If this course were persevered in for a succession of years, the other substances which would have been used in much more than the quantity required by the crops, must either have been entirely lost or have accumulated in the soil. In the latter case it is sufficiently obvious that the soil must have been gradually acquiring an amount of resources which must remain dormant until the system of manuring is changed. To render them available, it is only necessary to add to it a quantity of the particular substance in which the manure hitherto employed has been deficient, so as to restore the lost balance, and enable the plant to make use of those which have been stored up within it. The substance so used is called a special manure; that containing all the constituents of the crop is a general manure.

The distinction of these two classes of manures is very important in a practical point of view, because a special manure is not by itself capable of maintaining the life of plants, but is only a means of bringing into use the natural and acquired resources of the soil. In place of preventing or retarding its exhaustion, it rather accelerates it by causing the increased crops to consume more abundantly, and within a shorter period of time, those substances which it contains. On the other hand, a general manure prevents or diminishes the consumption of the elements of plant-food contained in the soil, and if added in sufficient abundance, may cause them to accumulate in it, and even enable an almost absolutely barren soil to yield a tolerable crop. General manures must therefore always be the most important and essential, and no others would be used if it were possible to obtain them of a composition exactly suited to the requirements of the crop to be raised. Practically, however, this condition cannot be fulfilled, because all the substances available for the purpose, and particularly farm-yard manure, are refuse matters, the exact composition of which is not under our control, and they do not necessarily contain their constituents either in the most suitable proportions, or the most available forms, and consequently when they are used during a succession of years, certain of their constituents may accumulate in the soil, and it is under such circumstances that special manures are both necessary and advantageous.

Several different substances, but more especially farm-yard manure, fulfil in a very remarkable manner the conditions of a general manure, and supply abundantly, not merely the mineral, but also the carbonaceous and nitrogenous matters necessary for building up the organic part of the plant; and hence its use is governed by principles of comparative simplicity, and really resolves itself into determining the best mode of managing it so as effectually to preserve its useful constituents, and, at the same time, to bring them into those forms of combination in which they are most available to the plant. But the employment of a special manure opens up nice questions as to the relative importance of the different elements of plants which have given rise to much controversy and difference of opinion.

In treating of the food of plants, it has been already observed that the fixed or mineral constituents which are contained in their ash, are necessarily derived exclusively from the soil, but that the carbon, hydrogen, nitrogen, and oxygen, of which their organic part is composed, may be obtained either from that source or from the air. The important distinction which thus exists between these two classes of substances, has given rise to two different views regarding the theory of manures. Basing his views on the presence of the organic elements in the air, Liebig has maintained that it is unnecessary to supply them in the manure, while others, among whom Messrs. Lawes and Gilbert have taken a prominent position, hold that, as a rule, fertile soils, cultivated in the ordinary manner, contain a sufficient supply of mineral matters for the production of the largest possible crops, but that the quantity of ammonia and nitric acid which the plants are capable of extracting from the air is insufficient, and must be supplemented by manures containing them. A large number of experiments have been made in support of these views, but the inferences which can be drawn from them are not absolutely conclusive on either side, and it is necessary to consider the matter in a general point of view.

Setting out from the proposition already so frequently referred to, that the plant cannot grow unless it receives a supply of all its elements, it must be obvious that if, to a soil containing a sufficiency of mineral matters to raise a given number of crops, a supply of ammonia be added, its total productive capacity cannot be thus increased; and though it may yield larger crops than it would have done without that substance, this can only be accomplished by a proportionate diminution of their number. In either case, the same quantity of vegetable matter will be produced, but the time within which it is obtained will be regulated by the supply of ammonia. That substance differs in no respect from any other element of plant-food, and used in this way is to all intents and purposes a special manure, and acts merely by bringing into play those substances which the soil already contains. Its effect may not be apparent until after the lapse of a very long period of time, but it ultimately leads to the exhaustion of the soil. If, on the other hand, a soil be continuously cropped until it ceases to yield any produce, it is manifest that the exhaustion must in this instance be entirely due to the removal of its available mineral nutriment, because the superincumbent air constantly changed by the winds must continue to afford the same unvarying supply of the organic elements, and the power of supporting vegetation would be restored to it, by adding the necessary inorganic matters. Hence when a soil, which in its natural condition is capable of yielding a certain amount of vegetable matter, is rendered barren by the removal of the crop, it may be laid down as an incontrovertible position, that its infertility is due to the loss of mineral matters, and that it may be restored to its pristine condition by the use of them, and of them only.

But the case is materially altered when we come to consider the course of events in a cultivated soil. The object of agriculture is to cause the soil, by appropriate treatment, to yield much more than its normal produce, and the question is, how this can be best and most economically effected in practice. According to Liebig, it is attained by adding to the soil a liberal supply of those mineral substances required by the plant, and that it is unnecessary to use any of the organic elements, because they are supplied by the air in sufficient quantity to meet the requirements of the most abundant crops. Other chemists and vegetable physiologists again hold that though a certain increase may be obtained in this way, a point is soon reached beyond which mineral matters will not cause the plant to absorb more ammonia from the air, although a further increase may be obtained by the addition of nitrogen in that or some other available form.

It is admitted on both sides, that all the elements of plant food are equally essential, and the controversy really lies in determining what practically limits the crop producible on any soil. The point at issue may be put in a clear point of view by considering the course of events on a soil altogether devoid of the elements of plants. If a small quantity of mineral matters be added to such a soil, it immediately becomes capable of supporting a certain amount of vegetation, deriving from the air the organic elements necessary for this purpose, and with every increase of the former, the air will be laid under a larger contribution of the latter, to support the increased growth, and this must proceed until the limit of supply from the atmosphere is reached. At this point a further supply of mineral matters alone must obviously be incapable of again increasing the crop, and it would thus be absolutely necessary to conjoin them with a proportionate quantity of organic substances. Liebig maintains that this limit is never attained in practice, but that the air affords ammonia and the other organic elements in excess of the requirements of the largest crop, while mineral matters are generally though not invariably present in the soil in insufficient quantity. Messrs. Lawes and Gilbert, on the other hand, believe that the soil generally contains an excess of mineral matters, and that a manure which is to bring out their full effect must contain ammonia, or some other nitrogenous substance fitted to supplement the deficient supply afforded by the atmosphere. In short, the question at issue is, whether there is or is not a sufficiency of atmospheric food to meet the demands of the largest crop which can practically be produced.

An absolutely conclusive reply to this question is by no means easy. The experiments by which it is to be resolved are complicated by the fact, that all soils capable of supporting anything like a crop, contain not only the mineral, but the organic elements of its food in large and generally in greatly superabundant quantity, and it is impossible satisfactorily to ascertain how much is derived from this source, and how much from the atmosphere. There are in fact no experiments in which the effects of a purely mineral soil have been ascertained. The important and carefully performed researches of Messrs. Lawes and Gilbert were made upon a soil which had been long under cultivation, and contained decaying vegetable matters in sufficient abundance to supply nitrogen to many successive crops, and it would be most unreasonable to assert that the produce they did obtain by means of mineral manures, drew the whole of its nitrogen from the air. On the contrary, it may be fairly assumed that the soil did yield a certain quantity of its nitrogenous compounds, but to what extent this occurs, it is impossible to determine. This difficulty is encountered more or less in all the other experiments, and precludes absolute conclusions. The same fallacy also besets the arguments of Liebig when he holds that the crop, increased by means of mineral manures alone, must derive the whole of the additional quantity of nitrogen which it contains from the air, as appears to be tacitly assumed throughout the whole discussion. So far from this being the case, it is just as likely that the mineral matters should cause the plants to take it from the soil, if it is there, as from the atmosphere.

Taking a general view of the whole question, it is evident that a certain amount of vegetation may always be produced by means of mineral manures, and the quantity obtained is generally much beyond the normal produce of the soil. But it is still open to doubt whether the largest possible crop can be thus obtained, although the balance of evidence is against it, and in favour of the addition of ammonia, and other nitrogenous and organic substances, to the soil. In actual practice manures containing nitrogen are more important, and more extensively applied than any others, and the quantity of that element thus used is very much larger than is generally supposed. Twenty tons of farm-yard manure, a quantity commonly applied, and often exceeded on well cultivated land, contain a sufficiency of organic matters to yield about 2-1/2 cwt. of nitrogen. A complete rotation, according to the six-course shift, contains almost exactly the same quantity of nitrogen, when we assume average crops throughout the whole, and it is thus made up.[K]

Lbs. of Nitrogen. 1. Turnips (13-1/2 tons) 60 2. { Wheat (28 bushels at 60 lbs.) 29 { Straw 16 3. Hay (2-1/2 tons) 56 4. { Oats (34 bushels at 40 lbs.) 27 { Straw 14 5. Potatoes (3 tons) 27 6. Wheat and straw as before 45 ---- Total 274

The supply is therefore quite sufficient for the requirements of the crop; and when it is borne in mind that a considerable quantity of ammonia and nitric acid is annually carried down by the rain, and that during a long rotation other substances are very generally used in addition to farm-yard manure, it is obvious that the crop need not depend to any extent upon what it derives from the air. What is true of the nitrogenous matters applies with still greater force to the mineral constituents of the manure. Twenty tons of farm-yard manure contain 32 cwt. of mineral matters, while the average crops of a six course-shift contain only 1088 lbs., or less than one-third of this quantity. It is obvious, therefore, that in well manured land there must be a gradual increase of all the constituents of plants, but that of the mineral matters is relatively much greater than that of the nitrogenous. If therefore from any cause the crop produced on a soil to which farm-yard manure had been applied were greatly to exceed the average, the amount of produce, so far as the soil is concerned, would be limited not by deficiency of mineral, but of nitrogenous food. Hence also when farm-yard manure is liberally applied, there is a gradual accumulation of valuable matters, and a progressive improvement of the productive capacity of the soil.

It is far otherwise, however, if a special manure is employed, because in that case the crop is thrown upon the resources of the soil itself for all its constituents except those contained in the substance employed, and by persisting in its exclusive use exhaustion is the inevitable result. It would be wrong, however, to infer from this, that special manures are to be avoided. On the contrary, great benefits are derived from their judicious employment, and the circumstances under which they are admissible may be readily gathered from what has already been said. They are agents which bring into useful activity the dormant resources of the soil, they restore the proper balance between its different constituents, and supply the excessive demand of some particular elements. Thus, for instance, in a soil containing an abundant supply of mineral matters, a salt of ammonia or nitric acid increases the crop, by promoting the absorption of the substances already present. So likewise a soil on which young cattle and milch cows have been long pastured has its fertility restored by phosphate of lime, because that substance is removed in the bones and milk in relatively much larger proportion than any others.

The choice of a special manure is necessarily dependent on a great variety of circumstances, and is governed partly by the nature of the soil, and partly by that of the crop. It is obvious that cases may occur in which any individual element of the plant may be deficient, and ought to be supplied, but experience has shown that, as a rule, nitrogen and phosphoric acid are the substances which it is most necessary to furnish in this way, and which in all but exceptional cases produce a marked effect on the crop. The other substances, such as potash, soda, magnesia, etc., occasionally act beneficially, but the results obtained from them are very uncertain, and frequently entirely negative.

It has been commonly asserted that phosphates are specially adapted to root crops, and ammonia or nitrates to the cereals, and this statement is so far true, that the former are used with advantage on the turnip, while the latter act with great benefit on grain crops and more especially on oats and barley. The effect of the latter, however, is more or less apparent in all crops and on all soils, because it promotes the assimilation of the mineral matters already present. But its peculiar importance lies in the power which it has of promoting the rapid development of the young plant, causing it to send its roots out into the soil, and to spread its leaves into the air, thus enabling it to take from those two sources, abundance of the useful substances existing in them. But it ought to be distinctly understood, that the statement that particular manures are specially suited to particular crops must be assumed with some reservation, because everything depends upon the nature of the food contained in the soil. It is well known that there are many soils in which ammonia acts more favourably on the turnip than phosphates, and vice versa, and the difference is often due to the previous treatment. In many cases in which ammonia when first used proved most beneficial, it now begins to lose its effect, and the reason no doubt is, that by its means the phosphates existing in these soils have been reduced in amount, while the ammonia has accumulated, so that a change in the system of manuring becomes necessary. A general manure may be used year after year in a perfectly routine manner, but where a special manure is employed, the importance of watching its effects, and altering it as circumstances indicate, cannot be over-estimated. The length of time during which special manures have been extensively used has not been sufficient to bring this prominently before the agriculturist, but its importance must sooner or later force itself upon him, and he will then see the necessity for studying the succession of manures as well as that of crops.

Hitherto we have considered a manure merely as a source from which plants derive their food, but it exercises a scarcely less important action on the chemical and physical properties of the soil. Farm-yard manure, which, as we shall afterwards see, contains a large amount of decomposing vegetable and animal matters, yields a supply of carbonic acid, which operates on the mineral constituents, promotes their further disintegration, and thus liberates their useful elements. It affects also their physical properties, for it diminishes the tenacity of heavy clays; each straw as it decomposes forming a channel through which the roots of plants, air, and moisture can penetrate more readily than through the stiff clay itself. On the other hand, it diminishes the porosity of light sandy soils, causes them to retain moisture, and generally makes their texture more suitable to the plant. Special manures probably act to some extent chemically on the soil, but the nature of the changes they produce is as yet imperfectly understood. Superphosphates which are highly acid in all probability act powerfully on the mineral substances, and common salt, which, though of little importance to the plant, occasionally produces very striking effects, appears to exercise some decomposing action on the soil. It is difficult, however, to trace the mode in which they operate on a substance of such complexity as the soil. Lime, as we shall afterwards see, acts by promoting the decomposition of the vegetable matters on the soil, and possibly some other substances may have a similar effect.

In the application of manures to the soil there are several circumstances which must be taken into consideration. It is generally stated that they ought to be distributed as uniformly as possible, but this is not always necessary nor even advisable, and certainly is not acted on in practice. Much must depend upon the nature both of crop and soil. When the former throws out long and widely penetrating roots, the more uniformly the manure is distributed the better; but if the rootlets are short, it is clearly more advisable that it should be deposited at no great distance from the seed. Practically this is observed in the case of the potato and turnip, which are short rooted, and where the manure is generally deposited close to the seed. But this course is never adopted with the long rooted cereals, the manure being usually applied to the previous crop, so that the repeated ploughings to which the soil is subjected in the interval may distribute what remains as widely and uniformly as possible. In soils which are either excessively tenacious or light, the accumulation of the manure close to the plants has also the effect of producing an artificial soil in their immediate neighbourhood, containing abundance of plant-food, and having physical properties better fitted for the support of the plant. On the other hand, when a special manure is used alone, and with the view of promoting the assimilation of substances already existing in the soil, the more uniform its distribution the better, because it is essential that the roots which penetrate through it should find at every point they reach not only the original soil constituents, but also the substances used to supplement their deficiencies.

FOOTNOTES:

[Footnote K: The quantities here taken are the averages deduced from the agricultural statistics taken in Scotland some years since, with the exception of hay and straw, which are not included in them. I have therefore assumed a reasonable quantity in these cases.]

CHAPTER VIII.

THE COMPOSITION AND PROPERTIES OF FARM-YARD AND LIQUID MANURES.

In the preceding chapter, a general manure has been defined as one containing all the constituents of the crop to which it is to be applied, in a state fitted for assimilation. This condition is fulfilled only by substances derived from the vegetable and animal kingdoms, and most effectually by a mixture of both. On this account, and also because its properties are such as enable it to act powerfully on the soil, farm-yard manure must always be of the highest importance. It is, in fact, the typical manure, and in proportion as other substances approach it in properties and composition, is their value for general purposes on the farm.

Farm-yard manure is a mixture of the dung and urine of domestic animals, with the straw used as litter; and its value and composition must necessarily depend upon that of these substances, as well as on the proportion in which they are mixed. The dung of animals consists of that part of their food which passes through the intestinal canal without undergoing assimilation; the urine containing the portion which has been assimilated and is again excreted, in consequence of the changes which are proceeding in the tissues of the animal. Their composition is naturally very different, and must be separately considered.

Urine.--Urine consists of a variety of earthy and alkaline salts, and of certain organic substances, generally rich in nitrogen, dissolved in a large quantity of water. That of the different domestic animals has been frequently examined, but the analyses of Fromberg give the most complete view of their manurial value:--

Horse. Swine. Ox. Goat. Sheep.

Extractive matter } 2?32 0?42 2?48 0?00 0?40 soluble in water } Extractive matter } 2?50 0?87 1?21 0?54 3?30 soluble in spirit} Salts soluble in } 2?40 0?09 2?42 0?50 1?57 water } Salts insoluble in} 1?80 0?88 0?55 0?80 0?52 water } Urea 1?44 0?73 1?76 0?78 1?62 Hippuric acid 1?60 ... 0?50 0?25 ... Mucus 0?05 0?05 0?07 0?06 0?25 Water 88?89 98?96 91?01 98?07 92?97 ---- -- ------- ------- ------- ------- 100?00 100?00 100?00 100?00 99?63

Composition of the Ash of these Urines.

Horse. Swine. Ox. Goat. Sheep. Carbonate of lime 12?0 ... 1?7 trace 0?2 Carbonate of magnesia 9?6 ... 6?3 7? 0?6 Carbonate of potash 46?9 12?0 77?8 trace ... Carbonate of soda 10?3 ... ... 53? 42?5 Sulphate of potash ... ... 13?0 ... 2?8 Sulphate of soda 13?4 7?0 ... 25? 7?2 Phosphate of soda ... 19?0 ... ... Phosphate of lime } Phosphate of magnesia } ... 8?0 ... ... 0?0 Chloride of sodium 6?4 53?0 0?0 14? 32?1 Chloride of potassium ... trace ... ... 12?0 Silica 0?5 ... 0?5 ... 1?6 Oxide of iron and loss 1?9 ... 0?7 ... ... ------ ------ ------ ------ --- --- 100?0 100?0 100?0 100?0 100?0

Human urine has been accurately examined by Berzelius, although his estimate of the proportion of urea is generally admitted to be above the average. His analysis gives the following numbers:--

Natural. Dry Residue. Urea 3?10 44?0 Lactic acid, lactate of ammonia,} 1?14 25?8 and extractive matter } Uric acid 0?00 1?9 Mucus 0?32 0?8 Sulphate of potash 0?71 5?4 Sulphate of soda 0?16 4?2 Phosphate of soda 0?94 4?9 Biphosphate of ammonia 0?65 2?6 Chloride of sodium 0?45 6?4 Muriate of ammonia 0?50 2?6 Phosphates of magnesia and lime 0?00 1?9 Silica 0?03 0?5 Water 93?00 ------- ------ 100?00 100?0

Among the special organic constituents of the urine are three substances, urea, uric acid, and hippuric acid, which are of much importance in a manurial point of view. The first of these is found in considerable quantity in the urine of all animals, but is especially abundant in the carnivora. Uric acid is found only in these animals, and is the most remarkable constituent of the excrement of birds, serpents, and many of the lower animals. Hippuric acid is most abundant in the herbivora. These substances are all highly nitrogenous. They contain--

	Urea.	Uric Acid.	Hippuric Acid.
Carbon	20·0	36?	60?
Hydrogen	6·0	2?	5?
Nitrogen	46·0	33?	8?
Oxygen	26·0	28?	26?
	100·0	100?	100?

They are extremely prone to change, and in presence of animal matters readily ferment, and are converted into salts of ammonia. Thus human urine, which, at the time of emission is free from smell of ammonia, and has a slightly acid reaction, becomes highly ammoniacal if it be kept for a few days. This is due to the conversion of urea into carbonate of ammonia; and the same change takes place, though more slowly, with uric and hippuric acids.

It is obvious, from the foregoing analyses, that great differences must exist in the manurial value of the urine of different animals. Not only do they vary greatly in the proportion of solid matters which they contain, but also in the kind and quantity of their nitrogenous constituents. They differ also in regard to their saline ingredients; and while salts of potash and soda form the principal part of the ash of the urine of the ox, sheep, goat, and horse, and phosphoric acid and phosphates are entirely absent, that of the pig contains a considerable quantity of the latter substances, and in this respect more nearly resembles the urine of man. Human urine is also much richer in urea and nitrogenous constituents generally, and has a higher value than any of the others.

It is especially worthy of notice that the urine of the purely herbivorous animals (with the exception of the sheep, which contains a small quantity), are devoid of phosphates and urea; and consequently, when employed alone, they are not general manures--a matter of some importance in relation to the subject of liquid manuring, which will be afterwards discussed.

Dung.--The solid excrement of animals is equally variable in composition. That of the domestic animals which had the ordinary winter food was found to have the following composition:--

	Horse.	Cow.	Sheep.	Swine.
Per-centage of water in the fresh excrement	77·5	82·5	56·7	77·3
Ash in the dry excrement	13·6	15·3	13·9	37·7

100 parts of ash contained--

Horse. Cow. Sheep. Swine. Silica 62?0 62?4 50?1 13?9 Potash 11?0 2?1 8?2 3?0 Soda 1?8 0?8 3?8 3?4 Chloride of sodium 0?3 0?3 0?4 0?9 Phosphate of iron 2?3 8?3 3?8 10?5 Lime 4?3 5?1 18?5 2?3 Magnesia 3?4 11?7 5?5 2?4 Phosphoric Acid 8?3 4?5 7?2 0?1 Sulphuric acid 1?3 1?7 2?9 0?0 Carbonic acid ... trace trace 0?0 Oxide of manganese 2?3 ... ... Sand ... ... 61?7 ---- - --- ---- ---- 99?0 99?9 99?4 99?2

Human f鎋es contain about 75 per cent of water; and their dry residue was found by Way to have the following composition:--

Organic matter 88?2 Insoluble siliceous matters 1?8 Oxide of iron 0?4 Lime 1?2 Magnesia 1?5 Phosphoric acid 4?7 Sulphuric acid 0?4 Potash 1?9 Soda 0?1 Chloride of sodium 0?8 ------ 100?0

In a sample analyzed by myself there were found--

Organic matter 86?5 Phosphates 8?9 Alkaline salts, containing 1?8 of phosphoric } 2?3 acid } Insoluble matters 2?3 ------ 100?0

Nitrogen 4?9 Equal to ammonia 5?7

It is to be observed that the urine and dung of animals differ conspicuously in the composition of their ash, the former being characterized by the abundance of alkaline salts, while the latter contains these substances in small proportion, but is rich in earthy matters, and especially in phosphoric acid. Salts of potash, for example, form nine-tenths of the inorganic part of the urine of the ox, while less than three per cent of that alkali is found in its dung. Phosphoric acid, on the other hand, is not met with in the urine, but forms about ten per cent of the dung. Silica is the most abundant constituent of the dung, but a large proportion of that found on analysis has been swallowed in the shape of grains of sand and particles of soil mechanically mixed with the food, although part is also derived from the straw and grains, which contain that substance in great abundance. The difference in the quantity of nitrogen they contain is also very marked, and is distinctly shown by the following analyses by Boussingault, which give the quantity of carbon, hydrogen, nitrogen, oxygen, and ash in the dung and urine of the horse and the cow in their natural state, and after drying at 212?

	HORSE.				COW.			
	Natural.		Dry.		Natural.		Dry.	
	Dung.	Urine.	Dung.	Urine.	Dung.	Urine.	Dung.	Urine.
Carbon	4?6	9?6	36?	38?	3?8	4?2	27?	42?
Hydrogen	0?7	1?6	3?	5?	0?0	0?9	2?	5?
Nitrogen	1?5	0?4	12?	2?	0?4	0?2	3?	2?
Oxygen	1?0	9?1	11?	37?	3?9	3?4	26?	37?
Ash	4?1	4?2	36?	16?	4?8	1?3	40?	12?
Water	87?1	75?1	0?	0?	88?1	90?0	0?	0?
	100?0	100?0	100?	100?	100?0	100?0	100?	100?

Hence, weight for weight, the urine of the horse, in its natural state, contains three times as much nitrogen as its dung; that of the cow twice as much; and the difference, especially in the horse, is still more conspicuous when they are dry.

It is obvious that the quality of farm-yard manure must depend--1. On the kind of animal from which it is produced; 2. On the quantity of straw which has been used as litter; 3. On the nature of the food with which the animals have been supplied; 4. On the extent to which its valuable constituents have been rendered available by the treatment to which it has been subjected; and 5. On the care which has been taken to prevent the escape of the urine, or of the ammonia produced by its decomposition.

The composition of farm-yard manure has engaged the attention of several chemists; but there are still many points on which our information regarding it is less complete than might be desired. Its investigation is surrounded with peculiar difficulties, not merely on account of its complexity, but because its properties render it exceedingly difficult to obtain a sample which fairly represents its average composition. In the case of long dung, these difficulties are so great that it is scarcely possible to overcome them; and hence, discrepancies are occasionally to be met with in the analyses of the most careful experimenters. The most minute and careful analyses yet made are those of Voelcker, who has compared the composition of fresh and rotten dung, and studied the changes which the former undergoes when preserved in different ways. He employed in his experiments both fresh and rotten dung, and subjected them to different methods of treatment. His analyses are given in the accompanying table, in which column 1 gives the composition of fresh long dung, composed of cow and pig dung. 2. Is dung of the same kind, after having lain in a heap against a wall, but otherwise unprotected from the weather for three months and eleven days in winter, during which time little rain fell. 3. The same manure, kept for the same time under a shed. 4. Well rotten dung, which had been kept in the manure heap upwards of six months. 5. The same, after having lain against a wall for two months and nine days longer.

	1	2	3
Water	66·7	69·6	67·2
Soluble organic matters	2·8	3·6	2·3
Soluble inorganic matters—			
Silica	0·37	0·79	0·39
Phosphate of lime	0·99	0·00	0·31
Lime	0·66	0·48	0·56
Magnesia	0·11	0·19	0·04
Potash	0·73	1·96	0·76
Soda	0·51	0·87	0·92
Chloride of sodium	0·30	0·06	0·58
Sulphuric acid	0·55	0·60	0·19
Carbonic acid and loss	0·18	0·75	0·45
	----- 4·40	----- 5·70	----- 4·20
Insoluble organic matters	25·6	18·4	20·6
Insoluble inorganic matters—			
Soluble silica	0·67	0·12	1·93
Insoluble silica	0·61	0·57	1·75
Oxide of iron, alumina, and phosphates	0·96	0·10	1·35

	4	5
Water	75·2	73·0
Soluble organic matters	3·1	2·0
Soluble inorganic matters—		
Silica	0·54	0·47
Phosphate of lime	0·82	0·29
Lime	0·17	0·18
Magnesia	0·47	0·18
Potash	0·46	0·60
Soda	0·23	0·82
Chloride of sodium	0·37	0·52
Sulphuric acid	0·58	0·72
Carbonic acid and loss	0·06	0·84
	----- 1·7	----- 2·6
Insoluble organic matters	12·2	14·9
Insoluble inorganic matters—		
Soluble silica	1·24	1·0
Insoluble silica	1·10	1·4
Oxide of iron, alumina, and phosphates	0·47	0·7

	1	2	3
Containing phosphoric acid	(0·78)	(0·77)	(0·98)
Equal to bone earth	(0·86)	(0·77)	(0·46)
Lime	1·20	1·91	1·68
Magnesia	0·43	1·29	

| | | Potash | 0?78 | 0?99 | 0?27 | 0?08 | | Soda | 0?19 | 0?46 | 0?38 | | Sulphuric acid | 0?61 | 0?99 | 0?98 | | Carbonic acid and loss | 0?84 | 0?29 | 1?77 | | | ----- 4?5 | ----- 4?0 | ----- 7?7 | | | ----- | ----- | ------ | | | 100?0 | 100?0 | 100?0 | | | | | | | Containing nitrogen | 0?49 | 0?70 | 0?70 | | Equal to ammonia | 0?81 | 0?20 | 0?06 | | Containing nitrogen | 0?94 | 0?70 | 0?80 | | Equal to ammonia | 0?99 | 0?70 | 0?00 | | Total nitrogen | 0?43 | 0?40 | 0?50 | | Equal to ammonia | 0?80 | 0?90 | 0?06 |

+-------------------------------+--------------+--------------+
| | 4 | 5 |
+-------------------------------+--------------+--------------+
Containing phosphoric acid	(0?74)	(0?6)
Equal to bone earth	(0?73)	(0?0)
Lime	1?67	2?5
Magnesia	0?91	0?2
Potash	0?45	0?2
Soda	0?38	0?1
Sulphuric acid	0?63	0?0
Carbonic acid and loss	1?95	1?4
	----- 6?8	---- 6?5
	-----	-----
	100?0	100?0
Containing nitrogen	0?97	0?49
Equal to ammonia	0?60	0?80
Containing nitrogen	0?09	0?13
Equal to ammonia	0?75	0?44
Total nitrogen	0?06	0?62
Equal to ammonia	0?35	0?24
+-------------------------------+--------------+--------------+

On examining and comparing these analyses, it appears that the differences are by no means great, although, on the whole, they tend to show that, weight for weight, well-rotten dung is superior to fresh, provided it has been properly treated. Not only is the quantity of valuable matters existing in the soluble state materially increased, whereby the dung is enabled to act with greater rapidity, but, owing to evaporation and the escape of carbonic acid, produced by the decomposition of the organic substances, the proportion of those constituents which are most important to the plant is increased. This is particularly to be noticed, in regard to the nitrogen, which has distinctly increased in all cases in which the dung has been kept for some time; and the practical importance of this observation is very great, because it has been commonly supposed that, during the process of fermentation, ammonia is liable to escape into the air. It would appear, however, that there is but little risk of loss in this way, so long as the dung-heap is left undisturbed; and it is only when it is turned that any appreciable quantity of ammonia volatilizes. It is different, however, with the action of rain, which soon removes, by solution, a considerable quantity of the nitrogen contained in farm-yard manure; and the deterioration must necessarily be most conspicuous in rotten dung, which sometimes contains nearly half of its nitrogen in a soluble condition. The effect produced in this way is conspicuously seen, by the results of weighings and analyses of small experimental dung-heaps, made by Dr. Voelcker at different periods. The subjoined table shows the composition of the heap, lying against a wall, and exposed to the weather at different periods:--

	WHEN PUT UP.			
	Nov 3d 1854.	April 30th 1855.	Aug 23d 1855.	Nov 15th 1855.
Weight of manure in lbs.	2838	2026	1994	1974
Water	1877?	1336?	1505?	1466?
Dry Matter	960?	689?	488?	507?
Consisting of--				
Soluble organic matter	70?8	86?1	58?3	54?4
" mineral matter	43?1	57?8	39?6	36?9
Insoluble organic matter	731?7	389?4	243?2	214?2
" mineral matter	114?4	155?7	147?9	201?7
Total nitrogen	18?3	18?4	13?4	13?3
Equal to ammonia	22?4	22?2	15?6	15?5

In this case, during the winter six months, which were very dry, the manure lost 541? lbs. of water and 270? lbs. of dry matter, but the nitrogen remained completely unchanged. But during the succeeding semi-annual period, when rain fell abundantly, the quantity of nitrogen is diminished by nearly a third, while the water has increased, and the loss of dry matter by fermentation, notwithstanding the high temperature of the summer months, was only 182? lbs. The soluble mineral matters also, which increased during the first period, are again reduced during the second, until they also fall to about two-thirds of their maximum quantity. That this effect is to be attributed to the solvent action of rain is sufficiently obvious, from a comparison of the results afforded by the other heaps, which had been kept under cover during the same period, as shown below.

	WHEN PUT UP.			
	Nov. 3d, 1854.	April 30th 1855.	Aug. 23d 1855.	Nov. 15th 1855.
Weight of manure in lbs.	3258	1613	1297	1235
Water	2156?	917?	563?	514?
Dry Matter	1102?	695?	733?	720?

Consisting of--				
Soluble organic matter	80?7	74?8	53?6	66?8
" mineral matter	50?4	54?1	39?5	54?8
Insoluble organic matter	839?7	410?4	337?2	341?7
" mineral matter	131?2	155?7	303?7	257?7
Total nitrogen	20?3	19?6	16?4	1?9
Equal to ammonia	25?0	23?3	20?8	2?1

The loss of nitrogen is here comparatively trifling, and during the whole year, but little exceeds two pounds, of which the greater part escapes during the first six months, and the soluble inorganic matters are almost unchanged. The total weight of the manure, however, undergoes a very great reduction, due chiefly to evaporation of water, but in part also to the loss of organic matters evolved in the form of carbonic acid during fermentation.

When the manure is spread out, as it is usually found under cattle in open yards, the deterioration is very great, a quantity thus treated having lost, in the course of a year, nearly two-thirds of its nitrogen, and four-fifths of its soluble inorganic matters.

The general conclusion deducible from these analyses is that, provided it be carefully prepared, farm-yard manure does not differ very largely in value, although the balance is in favour of the well-rotten dung. This result is in accordance with that obtained by other experimenters, who have generally found from 0? to 0? per cent of nitrogen, and 1 or 2 per cent of phosphates. But when carelessly managed, it may fall greatly short of this standard, as is particularly seen in a sample examined by Cameron, which had been so effectually washed out by the rain, as to retain only 0?5 per cent of ammonia. These cases, however, are exceptional, and well made and well preserved farm-yard manure will generally be found to differ comparatively little in value; and when bought at the ordinary price, the purchaser, as we shall afterwards more particularly see, is pretty sure to get full value for his money, and the specialities of its management are of comparatively little moment to him. But the case is very different when the person who uses the manure has also to manufacture it. The experiments already quoted have shown that, though the manure made in the ordinary manner may, weight for weight, be as valuable as at first, the loss during the period of its preservation is usually very large, and it becomes extremely important to determine the mode in which it may be reduced to the minimum.

In the production of farm-yard manure of the highest quality, the object to be held in view is to retain, as effectually as possible, all the valuable constituents of the dung and urine. But in considering the question here, it will be sufficient to refer exclusively to its nitrogen, both because it is the most important, and also because the circumstances which favour its preservation are most advantageous to the other constituents. In the management of the dung-heap, there are three things to be kept in view:-- First, To obtain a manure containing the largest possible amount of nitrogen; secondly, To convert that nitrogen more or less completely into ammonia; and thirdly, To retain it effectually.

As far as the first of these points is concerned, it must be obvious that much will depend on the nature and quantity of the food with which the animals yielding the dung are supplied, and the period of the fattening process at which it is collected. When lean beasts are put up to feed, they at first exhaust the food much more completely than they do when they are nearly fattened, and the manure produced is very inferior at first, and goes on gradually improving in quality as the animal becomes fat.

When the food is rich in nitrogenous compounds, the value of the manure is considerably increased. It has been ascertained, for instance, that when oil-cake has been used, not less than seven-eighths of the valuable matters contained in it reappear in the excrements; and as that substance is highly nitrogenous, the dung ought, weight for weight, to contain a larger amount of that element. That it actually does so, I satisfied myself by experiments, made some years since, when the dung and urine of animals fed on turnips, with and without oil-cake, were examined; but unfortunately, no determination of the total quantity of the excretions could be made, so that it was impossible to estimate the increased value. It has been commonly supposed that when cattle are fed with oil-cake, the increased value of the manure is equal to from one-half to two-thirds the price of the oil-cake; but this is a rather exaggerated estimate as regards linseed-cake, although it falls short of the truth in the case of rape, as we shall afterwards more particularly see.

Although it may be possible, in this way, to increase the quantity of nitrogen as a manure, there is a limit to its accumulation, due to the fact, that it is contained most abundantly in the urine, which can only be retained by the use of a sufficient supply of litter. Where that is deficient, the dung-heap becomes too moist, and the fluid and most valuable part drains off, either to be lost, or to be collected in the liquid manure-tank. In the well managed manure-heap, the quantity of litter should be sufficient to retain the greater part of the liquid manure, and to admit of only a small quantity draining from it, which should be pumped up at intervals, so as to keep the whole in a proper state of moisture. Attention to this point is of great moment, and materially affects the fermentation. When it is too moist or too dry, that process is equally checked; in the former case by the exclusion of air, which is essential to it; and in the latter, by the want of water, without which the air cannot act. The exact mode in which the manure is to be managed must greatly depend on whether the supply of litter is large or small. In the latter case the urine escapes, and is collected in the liquid manure-tank, and must be used by irrigation, and in some cases this mode of application has advantages, but in general, it is preferable to avoid it, and have recourse to substances which increase the bulk of the heap sufficiently to make it retain the whole of the liquid. For this purpose, clay, or still better, the vegetable refuse of the farm, such as weeds, ditch cleanings, leaves, and, in short, any porous matters, may be used. But by far the best substance, when it can be obtained, is dry peat, which not only absorbs the fluid, but fixes the ammonia, by converting it more or less completely into humate. Reference has been already made to the absorbent power of peat in the section on soils, but it may be mentioned here that accurate experiment has shown that a good peat will absorb about 2 per cent[L] of ammonia, and when dry will still retain from 1 to 1? per cent, or nearly twice as much as would be yielded by the whole nitrogen of an equal weight of farm-yard manure. Peat charcoal has been recommended for the same purpose, but careful experiment has shown that it does not absorb ammonia, although it removes putrid odour; and though it may be usefully employed when it is wished to deodorize the manure heap, it must not be trusted to for fixing the ammonia.

Much stress has frequently been laid on the advantage to be derived from the use of substances capable of combining chemically with the ammonia produced during the fermentation of dung and gypsum, sulphate of iron, chloride of manganese, sulphate of magnesia, and sulphuric acid, have been proposed for this purpose, and have been used occasionally, though not extensively. They all answer the purpose of fixing the ammonia, that is, of preventing its escaping into the air; but the risk of loss in this way appears to have been much exaggerated, for a delicate test-paper, held over a manure-heap, is not affected; and during fermentation, humic acid is produced in such abundance, as to combine with the greater part of the ammonia. The real source of deterioration is the escape of the soluble matters in the drainings from the manure-heap, which is not prevented by any of these substances; and where no means are taken to preserve or retain this portion, the loss is extremely large, and amounts, under ordinary circumstances, to from a third to a half of the whole value of the manure. Manure, therefore, cannot be exposed to the weather without losing a proportion of its valuable matters, depending upon the quantity of rain which falls upon it. Hence it is obvious that great advantage must be derived, especially in rainy districts, from covered manure-pits. This plan has been introduced on some farms with good effect; but the expense and doubts as to the benefits derived from it, have hitherto prevented the practice becoming general. The principal difficulty experienced in the use of the covered dung-pit is, that, where the litter is abundant, the urine does not supply a sufficiency of moisture to promote the active fermentation of the dung, and it becomes necessary to pump water over it at intervals; but when this is properly done, the quality of the manure is excellent, and its valuable matters are most thoroughly economized.

Although covered dung-pits have been but little used, their benefits have been indirectly obtained by the method of box-feeding, one of the great advantages of which is held to be the production of a manure of superior quality to that obtained in the old way. In box-feeding none of the dung or urine is removed from under the animals, but is trampled down by their feet, and new quantities of litter being constantly added, the whole is consolidated into a compact mass, by which the urine is entirely retained. Whatever objection may be taken to this system, so far as the health of the animals is concerned, there is no doubt as to the complete economy of the manure, provided the quantity of litter used be sufficient to retain the whole of the liquid. But its advantage is entirely dependent on the possibility of fulfilling this condition.

Whether box manure is really superior to that which can be prepared by the ordinary method is very questionable, but it undoubtedly surpasses a large proportion of that actually produced. It is more than probable, however, that the careful management of the manure-heap would yield an equally good product. It is manifest that the same number of cattle, fed in the same way, on the same food, and supplied with the same quantity of litter, must always excrete the same quantities of valuable matters; and the only question to be solved is, whether they are more effectually preserved in the one way than the other? It will be readily seen that this cannot be done by analysis alone, but that it is necessary to conjoin with it a determination of the total weight of manure produced; for though, weight for weight, box manure may be better than ordinary farm-yard manure, the total quantity obtained by the latter method, from a given number of cattle, may be so much greater, that the deficiency in quality may be compensated for. At the present time our knowledge is too limited to admit of a definite opinion on this subject, but it is highly deserving of the combined investigation of the farmer and the chemist.

Supposing the conditions which produce the manure containing the largest quantity of nitrogen to have been fulfilled, we have now to consider those which affect its evolution in the form of ammonia. This change is effected by fermentation. When a quantity of manure is left to itself it becomes hot, and gradually diminishes in bulk, and if it be turned over after some time, the smell of ammonia may be more or less distinctly observed. This ammonia is produced, in the first instance, from the urine, the nitrogenous constituents of which are rapidly decomposed, and the fermentation thus set up in the mass of manure extends first to the solid dung, and then to the straw of the litter, and gradually proceeds until a large quantity of ammonia is produced.

When fresh manure is deposited in the soil, the same changes occur, but they then proceed more slowly, and experience has shown that a much smaller effect is produced on the crop to which it has been applied than when it has been well fermented in the heap. This effect is consistent with theory, which would further indicate that well-fermented dung must be especially advantageous when applied to quick-growing crops, and less necessary to those which come slowly to maturity. As a rule, well fermented manure is to be preferred, provided it has been well managed and carefully prepared; but when this has not been done, and the manure has been exposed to the weather, or made in open courts or hammels, the economic advantages are all on the side of the fresh dung. It may be questioned also whether, now that there are so many other available sources of ammonia, it may not in many instances be advantageous to use the dung fresh, conjoined with a sufficient quantity of some salt of ammonia, or other substance fitted to supply the quantity of that element necessary for the requirements of the crop.

After the farm-yard manure has been prepared at the homestead, it is often necessary to cart it out to the field some time before it is to be applied, and it is a question of some importance to determine how it may be best preserved there. The general practice is to store it in heaps in the corners of the fields, but some difference of opinion exists as to whether it should be lightly thrown up so as to leave it in a porous state, and so promote its further fermentation, or whether it should be consolidated as much as possible by driving the carts on to the top of the heap during its construction. Considering the risks to which the manure is exposed on the field, the latter plan would appear to be the best. It is advisable also to interstratify the dung with dry soil, so as to absorb any liquid which may tend to escape from it, and it should also be covered with a well-beaten layer of earth, in order to exclude the rain. Although these precautions must not be omitted if the manure is to be stored in heaps, it will probably be often found quite as advantageous to spread it at once, and leave it lying on the surface until it is convenient to plough it. The loss of ammonia by volatilization will, under such circumstances, especially in the cold season of the year, be very trifling, and the rain which falls will only serve to incorporate the soluble matters with the soil, where they will be retained by its absorptive power.

In the actual application of the manure to the crop, several points require consideration. It is especially important to determine whether it ought to be uniformly distributed through the soil, or be kept near the roots of the plants. Both systems have their advocates, and each has advantages in particular cases. The choice between the two must greatly depend upon the nature of the crop and the soil. When the former is of a kind which spreads its roots wide and deep through the soil, the more uniformly the manure can be distributed the better; but when it is used with plants whose roots do not travel far, it is more advantageous to accumulate it near the seeds. Obvious advantages also attend this practice in soils which are either too heavy or too light. When, for example, it is necessary to cultivate turnips in a heavy clay, the manure put into the drills produces a kind of artificial soil in the neighbourhood of the plants, in which the bulbs expand more readily than in the clay itself. On the other hand, when a large quantity of dung, in a state of active fermentation, comes into immediate contact with the roots, its effect is not unfrequently injurious. These and many other points, which will readily suggest themselves to any one who studies the composition and properties of farm-yard manure, belong more strictly to the subject of practical agriculture, and need not be enlarged on here.

In the present state of agriculture, a proper estimate of the money value of farm-yard manure is of much importance in an economic point of view, and many matters connected with the profitable management of a farm must hinge upon it. If an estimate be made upon the principle which will be explained when we come to treat of artificial manures, it appears that fresh farm-yard manure of good quality is worth from 12s. to 15s. per ton, and well-rotted dung rather more. It is questionable, however, whether the system of valuation which is accurate in the case of a guano or other rapidly acting substance, is applicable to farm-yard manure, the effects of which extend over some years. A deduction must be made for the years during which the manure remains unproductive, and also for the additional expense incurred in carting and distributing a substance so much more bulky than the so-called portable manures, and it would not be safe to estimate its value at more than 7s. or 8s. per ton.

Liquid Manure.--This term is applied to the urine of the animals fed on the farm, and to the drainings from the manure-heap, which, in place of being returned to it, are allowed to flow away, and collected in tanks, from which they are distributed by a watering-cart, or according to the method recently introduced in Ayrshire, and since adopted in other places, by pipes laid under-ground in the fields, and through which the manure is either pumped by steam-power, or, where the necessary inclination can be obtained, is distributed by gravitation. That liquid manure must necessarily be valuable, is an inference which maybe at once drawn from the analyses of the urine of different animals already given, and of which it chiefly consists. In addition to the urine, however, it contains also the soluble organic and mineral matters of the dung, as well as a quantity of solid matters in suspension, among which phosphates are found, and thus it possesses a supply of an element which would be almost entirely deficient if it were composed of urine alone. In the following analyses by Professor Johnston, No. 1 is the drainings of the manure-heap when exposed to rain; and No. 2 the same, when moistened with cows' urine pumped over it, the results being expressed in grains per gallon:--

No. 1. No. 2. Ammonia 9? 21? Organic matter 200? 77? Ash 268? 518? ----- ----- Total solids in a gallon 479? 617?

The ash contained--

Alkaline salts 207? 420? Phosphates 25? 44? Carbonate of lime 18? 31? Carbonate of magnesia, and loss 4? 3? Silica and alumina 13? 19? ----- ----- 268? 518?

More elaborate analyses of the same fluid have since been made by Dr. Voelcker, with the subjoined results per gallon:--

1. 2. 3. Organic matters and ammoniacal } 263?0 250?3 70?21 salts } Silica 2?9 9?8 1?54 Oxide of iron 0?0 0?8 ... Lime 5?4 25?8 13?11 Magnesia 2?6 15?3 1?60 Potash 103?3 112?6 13?11 Chloride of potassium 72?0 77?8 7?12 Chloride of sodium 17?8 46?3 17?58 Phosphoric acid 2?0 9?1 2?04 Sulphuric acid 22?1 37?0 3?08 Carbonic acid, and loss 33?0 27?5 14?25 ------ ------ ------- Total solids 526?1 612?3 144?64 Ammonia 114?6 22?1 26?47

The differences are here very remarkable, especially in the quantity of ammonia, which is exceedingly large in the first sample. All of them are particularly rich in potash, and contain but a small proportion of phosphoric acid. The general inference to be deduced from them is, that liquid manure is a most important source of the alkalis and ammonia, and must be peculiarly valuable on soils in which these substances are deficient.

The system of liquid manuring, originally introduced by Mr. Kennedy of Myremill, Ayrshire, and which has since been adopted in some other places, differs from liquid manuring in its strict sense, for not only are the drainings of the manure-heap employed, but the whole solid excrements are mixed with water in a tank, and rape-dust and other substances occasionally added, and distributed through the pipes.

It has been abandoned on Mr. Kennedy's farm, but is in use at Tiptree Hall, and on the farm of Mr. Ralston, Lagg, where the fluid is distributed by gravitation.

The arrangements employed by Mr. Mechi are identical with those formerly in use at Myremill. The greater part of the stock is kept on boards, and the liquid and solid excrements are collected together in the tank, and largely diluted before distribution. The liquid from the tanks has been recently examined by Dr. Voelcker, who found it to contain per gallon--

Organic matter and ammoniacal salts 53·3 Soluble silica 6·7 Insoluble siliceous matter (clay) 15·7 Oxide of iron and alumina 2·6 Lime 6·0 Magnesia 1·3 ----- Potash 0·5 Chloride of potassium 1·5 Chloride of sodium 4·1 Phosphoric acid 3·2 Sulphuric acid 1·4 Carbonic acid, and loss 0·7 ----- Total solids 96·0 Ammonia 8·0

The quantity of this liquid distributed per acre is about 50,000 gallons, at a cost of 2d. per gallon. As this quantity contains about 39 lbs. of ammonia, it must be nearly equivalent to 2 cwt. of Peruvian guano, which costs, with the expense of spreading, from 28s. to 30s. per acre, while the cost of distributing the liquid exceeds ?: 17s. per acre. On the other hand, the rapidity with which liquid manure produces its effect must be taken into account. It is on this that its chief value depends, and especially when applied to grass land in early spring, it produces an abundant crop just when turnips and other winter food are exhausted. Mr. Telfer, Cunning Park, who has used this system for a good many years, has come to the conclusion that it is only in this way that it can be made profitable; and though pipes are laid all over his farm, he has latterly restricted the use of the liquid manure entirely to Italian ryegrass. Its effect on the cereals is much less marked, and it can scarcely be considered as capable of advantageous application to the general operations of the farm. Neither can liquid manure be applied to all soils. It fails entirely on heavy clays, but is peculiarly adapted to light sandy soils; and even barren sand may by its repeated application, be made to yield luxuriant crops. It is not likely that the system of liquid manuring will extend, except in localities where it is possible to distribute it by gravitation; and even then, it will probably be found most economical to restrict its use to one portion of the farm; and for that purpose, the poorest and most sandy soil ought to be selected.

Sewage Manure.--The use of the sewage of towns as a manure is closely connected with that of the liquid manure produced on the farm. Its application must take place in a similar manner, and be governed by the same principles. Although numerous attempts have been made to convert it into a solid form, or to precipitate its valuable matter, none of them have succeeded; nor can it be expected that any plan can be devised for the purpose, because the most important manurial constituents are chiefly soluble, and cannot be converted into an insoluble state, or precipitated from their solution. In its liquid form, however, sewage manure has been employed with the best possible effect in the cultivation of meadows. The most important instance of its application is in the neighbourhood of Edinburgh, where 325 acres receive the sewage of nearly half the town, and have been converted from barren sand into land which yields from ?0 to ?0 per acre. The contents of the sewer, taken just before it flows into the first irrigated meadow, near Lochend, were found to contain per gallon--

Soluble organic matter 21?0 Insoluble organic matter 21?0 Peroxide of iron and alumina 2?1 Lime 10?0 Magnesia 2?0 Sulphuric acid 6?9 Phosphoric acid 6?4 ----- Chlorine 12?0 Potash 2?9 Soda 13?7 Silica 6?0 ------ 105?0 Ammonia 14?0

It is interesting to notice that this sewage is superior in every respect to the liquid manure used at Tiptree Hall; and the good effects obtained from its application, in the large quantities in which it is used in the Craigentinny meadows, may be well imagined. It operates, not merely by the substances which it holds in solution, but also by depositing a large quantity of matters carried along in suspension, and is in reality warping with a substance greatly superior to river-mud. A deposit collected in a tank, where the sewage passes through a farm, is used as a manure, and contains--

Peroxide of iron and alumina 4?5 Lime 1?4 Magnesia 0?9 Potash 0?0 Soda 0?6 Phosphoric acid 1?8 Sulphuric acid 0?6 Organic matter 17?5 Sand 20?1 Water 53?6 ------ 100?0 Ammonia 0?3

And even, though containing more than half its weight of water and 20 per cent of sand, this substance has considerable value as a manure.

The growing evils of the existing system of sewage, and the enormous waste of a manurial matter, which the experience of the Craigentinny meadows has shewn to be productive of the most important effects, has recently directed much attention to the conversion of the contents of our sewers into a useful manure. Numerous plans for its precipitation and conversion into a solid manure have been proposed, but most of these have shewn an entire ignorance of the fundamental principles of chemistry, and the best only succeed in precipitating a very small proportion of its valuable matters, and leave almost the whole of the ammonia, as well as the greater part of the fixed alkalies, in solution. Nor is it to be expected that any process will be discovered by which these substances can be precipitated, because solubility is the special characteristic of their compounds, and no means is known by which it is possible to convert them into an insoluble form. If sewage is to be used at all, there seems little doubt that it must be by applying it entire, and in the liquid state. But here again, the expense of conveying it on to the land becomes an obstacle which it must frequently be impossible to overcome. When it can be conveyed by gravitation, as is the case in the neighbourhood of Edinburgh, it may undoubtedly be used with the utmost advantage, and with the very best economic results. But when it requires to be carried to a great distance through pipes, and raised to a high level by pumping, all these advantages disappear. If the cost of application amounts to 2d. a gallon, as in Mr. Mechi's case, or even to half that sum, it may be fairly concluded that it cannot be used with any great prospect of large economic results, and that, unless under very exceptional cases, it must be unprofitable.

The chances of success must also greatly depend upon the kind of soil on which it is used. Experience has shewn that its effects are most beneficial on light and deep sandy soils, but that on heavy retentive clays it is without effect, or even absolutely injurious. In clay soils it is important to use every means of getting rid of moisture, and any plan which adds 200 or 300 tons of water to them, only aggravates their natural defects to an extent which more than counterbalances the benefits derived from the manurial matter it contains. Whatever the ultimate result of the use of town sewage in the liquid form may be, it is unlikely that it will be employed in general agricultural practice. It is more probable that it will be found necessary to set apart a certain breadth of land to be treated by it exclusively. Many plans have been proposed for conveying it through considerable districts, and selling to the surrounding farmers the quantities which they require, but wherever large sewage-works are established, it will be impossible to depend on a precarious demand, and the promoters of such schemes will be compelled, as part of their speculation, to supply not only the manure, but the land on which it is to be used. Indeed, the difficulties attending the whole question are so formidable, that even those who are most anxious to see a stop put to the waste of manurial matter must admit that the prospect of a successful economic result is not encouraging. Nor is it likely that anything will be done until the whole system of managing town refuse is changed, and in place of deluging it with water, some plan can be contrived which, while fulfilling sanatory requirements, shall preserve it in a concentrated form, or convert it into a dry and inodorous substance.

FOOTNOTES:

[Footnote L: Report on the economic uses of peat. Highland Society's Transactions, N.S., vol. iv. p. 549.]

CHAPTER IX.

COMPOSITION AND PROPERTIES OF VEGETABLE MANURES.

Many vegetable substances have been employed as manures, either alone or as auxiliaries to farm-yard manure. Like that substance, they are general manures, and contain all the constituents of ordinary crops; but, owing to the absence of animal matter, they in general undergo decomposition and fermentation much more slowly, although some of them contain a so largely preponderating proportion of nitrogen, that they may in some respects be compared to the strictly nitrogenous manures.

Rape-dust, Mustard, Cotton and Castor Cake.--Rape-dust has long been employed as a manure, and the success which has attended its use has led to the introduction of the refuse cake from some other oil seeds, such as those of mustard and castor-oil, which cannot be employed for feeding. Like the seeds of all plants, these substances are rich in nitrogen, and their ash, containing of course all the constituents of the plant, supplies the necessary inorganic elements. The following are analyses of these substances, which, in addition to the amount of nitrogen and phosphates, shew also that of water and oil, to which reference will be made in a future chapter, in relation to the feeding value of some of them. The detailed composition of their ash may be judged of from that of the seeds from which they are made, and which have been given under that head.

	Rape-Cake.	Poppy-Cake.	Cotton-seed Cake.	Castor-Cake.
Water	10·8	11·3	11·9	12·1
Oil	11·0	5·5	9·8	24·2
Albuminous compounds }	29·3	31·6	25·6	21·1
Ash	7·9	12·8	5·4	6·8
Other constituents	40·0	38·8	48·3	35·8
	100·0	100·0	100·0	100·0
Nitrogen	4·8	4·4	3·5	3·0
Silica	1·8	3·6	1·2	1·6
Phosphates	3·7	6·9?	2·9	2·1
Phosphoric acid } in combination } with alkalies }	0·9	3·7	0·5	0·4

A general similarity may be observed in the composition of all these substances; they are rich in nitrogen, and contain as much of that element as is found in six or seven times their weight of farm-yard manure, and a somewhat similar proportion exists in the amount of phosphates, and probably of their other constituents. They have all been employed with success, but the most accurate observations have been made with rape-dust, which has been longer and more extensively used than any of the others. It has been employed alone for turnips, or mixed with farm-yard manure, and also as a top-dressing to cereals. But the most marked advantage is derived from it when applied in the latter way on land which has been much exhausted, and its effects are then very striking. An adequate supply of moisture is essential to the production of its full effects, and hence it often proves a failure in very dry seasons, and on dry soils. It must not be applied in too great abundance, experience having shewn that after a certain point has been reached, an increase in the quantity produces no benefit, and even sometimes positively diminishes the crop. The other substances of the same class, in all probability, act in the same way, but as their introduction is recent, and their use limited, less is known regarding their effects.

Malt-Dust, Bran, Chaff, etc.--The value of these substances as manures is chiefly dependent on the nitrogen they contain, though to some extent also on their inorganic constituents. Malt-dust contains about 4? per cent, and bran 3? per cent of nitrogen. But they are little used as manures, as they can generally be more advantageously employed for feeding. The value of chaff more nearly resembles that of straw.

Straw is occasionally employed as a manure, and sometimes even as a top-dressing for grass land. It is generally admitted, however, that its application in the dry state, and especially as a top-dressing, is a practice not to be recommended, as it decomposes too slowly in the soil; and it is always desirable to ferment it in the manure heap, so as to facilitate the production of ammonia from its nitrogen. Still circumstances may occur in which it becomes necessary to employ it in the dry state, and it will generally prove most valuable on heavy soils, which it serves to keep open, and so promotes the access of air, and enables it to act on the soil. On light sandy soils it generally proves less advantageous, as its tendency of course is to increase the openness of the soil, and render it less able to retain the essential constituents of the plant.

The quantity of nitrogen in straw does not exceed 0? per cent, and its value is mainly due to its inorganic constituents and to its mechanical effect on the soil.

Saw-dust has little value as a manure, as it undergoes decomposition with extreme slowness. It is a good mechanical addition to heavy soils, and diminishes their tenacity; and though its manurial effects are small, it sooner or later undergoes decomposition, and yields what valuable matters it contains. The saw-dust of hard wood is to be preferred, both because it contains more valuable matters than that of soft wood, and because the absence of resinous matters permits its more rapid decomposition. It is a useful absorbent of liquid manure, and may be advantageously added to the dung-heap for that purpose.

Manuring with Fresh Vegetable Matter--Green Manuring.--The term green manuring is applied to the system of sowing some rapidly growing plant, and ploughing it in when it has attained a certain size, and the success attending it, especially on soils poor in organic matters, is very marked. It is obvious that this mode of manuring can add nothing to the mineral matters contained in the soil, and its utility must therefore be due to the plant gathering organic matters from the air, which, by their decomposition, yield nitrogen and carbonic acid--the former to be directly made use of by subsequent crops, the latter, in all probability, acting also on the soil, and setting free its useful constituents. Hence those plants which obtain the largest quantity of their organic elements from the air ought to be most advantageous for green manuring. The plants used for this purpose act also as a means of bringing up from the lower parts of the soil the valuable matters which exist in it out of reach of ordinary crops, and mixing them again with the surface part. Many of the plants found most useful for green manuring send down their roots to a considerable depth; and when they are ploughed in, all the substances which they have brought up are of course deposited in the upper few inches of the soil. Vegetable matter when ploughed in in the fresh state, also decomposes rapidly, and is therefore able immediately to improve the subsequent crop; and as this decomposition takes place in the soil without the loss of ammonia and other valuable matters, which is liable to occur to a greater or less extent when they are fermented on the dung-heap, it will be obvious that in no other mode can equally good results be obtained by its use.

Many plants have been employed as green manure, and different opinions have been expressed as to their relative values. In the selection of any one for the purpose, that should of course be taken which grows most rapidly, and produces within a given time the largest quantity of valuable matters, but no general rule can be given for the selection, as the plant which fulfils those conditions best will differ in different soils and climates. The plants most commonly employed in this country are spurry, white mustard, and turnips. Rye, clover, buckwheat, white lupins, rape, borage, and some others, have been largely employed abroad. Some of these are obviously unfitted for the climate of the British Islands; and the others, although they have been tried occasionally, do not appear to have been very extensively employed. The turnip is sown broadcast at the end of harvest, and ploughed in after two months. White mustard and spurry are employed in the same way as a preparation for winter wheat, and with the best results. The latter is sometimes sown as a spring crop in March, ploughed in in May, and another crop sown which is ploughed in in June, and immediately followed by a third. The effect of this treatment is such that the worst sands may be made to bear a remunerative crop of rye.

It is not easy to estimate the addition made by green manuring to the valuable matters contained in the soil, but it is probably far from inconsiderable. A crop of turnips, cultivated on the ordinary agricultural system, after two months' growth, weighs between five and seven tons per acre, and contains nitrogen equivalent to about 48 lbs. of ammonia, and half a ton of organic matters; but nothing is known as to the quantity produced when it is sown broadcast, and is not thinned, although it must materially exceed this. Neither is it possible to determine the relative proportions derived from the soil and the air, although it is, in all probability, dependent on the resources of the soil itself,--plants grown on a rich soil obtaining their chief supplies from it, while, on poorer soils, a larger proportion is drawn from the atmosphere. Hence light and sandy soils are most benefited by green manuring, partly on this account, and partly also, no doubt, because the valuable inorganic matters, which are so liable to be washed out of these soils, are accumulated by the plants and retained in them in a state in which they are readily available for the subsequent crop.

Sea-Weed.--Sea-weeds have been employed from time immemorial as a manure on the coasts of Scotland and England, in quantities varying from 10 to 20 tons per acre. Their action is necessarily similar to that of green manure ploughed in, as they contain all the ordinary constituents of land plants.

The subjoined analyses of three of the most abundant species will sufficiently indicate their general composition.

	LAMINARIA DIGITATA. Frond Collected in Spring.	Mixed Weeds in the state in which they actually are used.	Fucus nodosus.	Fucus vesiculosus.	Fucus Stem and Frond collected in Autumn.	
Water	74?1	70?7	88?9	77?1	80?4	
Albuminous compounds	1?6	2?1	0?3	3?2	2?5	
Fibre, etc.	19?4	22?5	4?2	10?9	6?0	
Ash	4?9	5?7	5?6	8?8	10?1	
	100?0	100?0	100?0	100?0	100?0	
Nitrogen	0?8	0?2	0?5	0?3	0?5	

The ash consisted of

					Stem.	Frond.
Peroxide of iron	0?5	0?5	0?0	0?0	0?5	2?5
Lime	9?0	8?2	7?1	7?9	4?2	18?5
Magnesia	6?5	5?3	2?3	5?1	10?4	6?8
Potash	20?3	20?5	5?5	11?1	12?6	12?7
Chloride of potassium	...	...	58?2	26?9	25?3	9?0
Iodide of potassium	0?4	0?3	1?1	2?9	1?2	1?8
Soda	4?8	6?9	...	...	...	...
Sulphuret of sodium[M]	3?6	...	...	...	...	...
Chloride of sodium	24?3	24?1	15?9	30?7	19?4	22?8
Phosphoric acid	1?1	2?4	2?2	2?6	1?5	4?9
Sulphuric acid	21?7	28?1	2?3	8?0	7?6	6?2
Carbonic acid	6?9	2?0	4?1	2?9	15?3	13?8
Silicic acid	0?8	0?7	0?3	0?9		

The first four analyses give the composition of the weeds after they have been separated from all foreign substances; the last, that of the mixture taken from the heap just as it is used in Orkney; and its value is then enhanced by small shells and marine animals adhering to the plants, which increase the amount of phosphoric acid and nitrogen.

The ease with which all sea-weeds pass into a state of putrefaction, adapts them in a peculiar manner to the manurial requirements of a cold and damp climate. The rapidity of their decomposition is such, that when spread on the land they are seen to soften and disappear in a short time. They form therefore a rapid manure, and their effects are said to be confined to the crop to which they are applied; but this is probably due to the fact, that they are chiefly used in inferior sandy soils, in which any manure is rapidly exhausted. In good soils there is no reason why their effect should not be as lasting as that of farm-yard manure, which, in many particulars, they considerably resemble. The method of applying sea-weeds most generally in use, is to spread them on the soil, and plough them in after putrefaction has commenced, and it is on the whole the most advantageous. But they are sometimes composted with lime and earth, or mixed with farm-yard manure, and occasionally, also, they are used as a top-dressing to grass land.

On some parts of the western coast of Scotland and in the Hebrides, sea-weed is the chief manure. It gives excellent crops of potatoes, but they are said to be of inferior quality, unless marl or shell-sand is employed at the same time.

Leaves may be used as a manure, simply by ploughing them in, by composting them with lime, or by adding them to the manure heap.

Peat.--As a source of organic matter, peat may be used with advantage, especially on soils in which it is naturally deficient. Dry peat of good quality contains about one per cent of nitrogen, and a quantity of ash varying from five to twenty per cent. These substances, however, become available very slowly, owing to the tardy decay of peat in its natural state; and in order to make it useful, it is necessary to compost it with lime, or to mix it with farm-yard manure, or some readily putrescible substance, so that its decomposition may be accelerated. It may be most advantageously used as an absorbent of liquid manure, and on this account, forms a useful addition to the manure heap.

The observations which have been made regarding the use of these substances, lead directly to the inference that all vegetable matters possess a certain manurial value, and that they ought to be carefully collected and preserved. In fact, the careful farmer adds everything of the sort to his manure heap, where, by undergoing fermentation along with the manure, their nitrogen becomes immediately available to the plant; while the seeds of weeds are destroyed during the fermentation, and the risk of the land being rendered dirty by their springing up when the manure comes to be used is prevented.

FOOTNOTES:

[Footnote M: The presence of sulphuret of sodium in this case is due to the difficulty of completely burning the ash. It exists in the plant as sulphate of soda.]

CHAPTER X.

COMPOSITION AND PROPERTIES OF ANIMAL MANURES.

Manures of animal origin are generally characterized by the large quantity of nitrogen they contain, which causes them to undergo decomposition with great rapidity, and to yield the greater part of their valuable matters to the crop to which they are applied.

Guano.--By far the most important animal manure is guano, which is composed of the solid excrements of carnivorous birds in a more or less completely decomposed state, and is accumulated in immense quantities on the coasts of South America and other tropical countries. It has been used as a manure in Peru from time immemorial, but the accounts given by the older travellers of its marvellous effects were considered to be fabulous, until Humboldt, from personal observation, confirmed their statements. It was first imported into this country in 1840, in which year a few barrels of it were brought home; and from that time its importation rapidly increased. Soon after large deposits of it were found in Ichaboe; and it has since been brought from many other localities. The quantity of guanos of all kinds imported into this country and retained for home consumption now exceeds 240,000 tons a year.

The value of guano differs greatly according to the extent to which its decomposition has gone, and this is chiefly dependent on the climate of the locality from which it is obtained. When deposited in the rainless districts of Peru it still retains some of the uric acid and the greater part of the ammonia naturally existing in it, and the quantity which has escaped by decomposition is unimportant. But that obtained from other districts has suffered a more or less complete decomposition according to the humidity of the climate, which reduces the quantity of organic matters and ammonia, until, in some varieties, they are so small as to be of little importance. The following are minute analyses of three specimens of Peruvian guano, shewing all the different constituents it contains, and the amount of difference which may exist:--

	I.	II.	III.
Urate of ammonia	10?0	9?	3?4
Oxalate of ammonia	12?8	10?	13?5
Oxalate of lime	5?4	7?	16?6
Phosphate of ammonia	19?5	6?	6?5
Phosphate of magnesia and ammonia	...	2?	4?0
Sulphate of potash	4?0	5?	4?3
Sulphate of soda	1?5	3?	1?2
Sulphate of ammonia	3?6	...	...
Muriate of ammonia	4?1	4?	6?0
Phosphate of soda	...	...	5?9
Chloride of sodium	...	...	0?0
Phosphate of lime	15?6	14?	9?4
Carbonate of lime	1?0	...	...
Sand and alumina	1?9	4?	5?0
Water	9?4 }	} 32?	23?2
Undetermined humus-like organic matters	10?0 }	}	-
	100?8	100?	100?0

These analyses illustrate two points--first, that in some samples the decomposition has advanced to a greater extent than in others; for we observe that the quantity of uric acid, or rather of urate of ammonia, is greatly less in the last analysis than in the other two, and much smaller than in the fresh dung, which contains from 50 to 70 per cent of uric acid; and secondly, that guano is rich in all the constituents of the plant, but especially in ammonia, the best form in which nitrogen can be supplied, in uric acid which by decomposition yields ammonia, and in phosphoric acid. But such analyses are too elaborate for ordinary purposes, and much less convenient for comparison and for estimating the value of the guano than the shorter analysis commonly in use, which gives the water, the loss by ignition (that is, the sum of the organic matters and ammoniacal salts), the phosphates, the alkaline salts, and the quantity of phosphoric acid contained in them, and existing there in a state similar to that in which it is found in the soluble phosphates of a superphosphate. In addition to these, the quantities of sand and other less valuable ingredients are also stated.

In the subjoined tables the composition of a great variety of different kinds of guano is given. Most of these are averages deduced from a considerable number of analyses of good samples. Those of some kinds of guano, such as Peruvian, which present a considerable amount of uniformity, afford a sufficiently accurate idea of the general composition of the variety, but in other cases they are of less value, because the imports of different seasons, and even of different cargoes, differ so greatly in composition that no proper average can be made. Several of these varieties are already exhausted, the importation of others has ceased, and new varieties are constantly being introduced.

Table showing the Average Composition of different varieties of Guano.

```
-------------------+--------+-------+-------------+---------------------------+
|Angamos.|Peru-   | ICHABOE.  |Bolivian or Upper |   |   | vian. | |Peruvian. | | |
+--------------+------+-----------+--------+     |   |   |Old.   |New.        |Old.
|Government.|Inferior| --------------------+--------+-------+------+------+------+--------
---+--------+ Water | 12?6 | 13?3 |24?1 |18?9 | 12?5| 16?4 | 14?5 | Organic
matter }| | | | | | | and ammoniacal }| 59?2 | 53?6 |39?0 |32?9 | 35?9|
12?8 | 26?4 | salts }| | | | | | | Phosphates | 17?1 | 23?8 |30?0 |19?3 |
27?3| 56?9 | 23?3 | Sulphate of lime | ... | ... | ... | ... | ... | ... | 9?5 |
Carbonate of lime | ... | ... | ... | ... | ... | ... | 12?7 | Alkaline salts | 7?0 | 7?7
| 4?9 | 8?2 | 15?9| 11?3 | 5?7 | Sand | 3?1 | 1?6 | 2?0 | 6?2 | 8?4| 2?1 |
8?9 | +--------+-------+------+------+------+-----------+--------+ | 100?0 |100?0
|100?0|100?0|100?0|100?0 |100?0 | Ammonia | 21?0 | 17?0 | 8?0| 10?2|
8?9| 2?7 | 3?6 | Phosphoric }| | | | | | | acid in alkaline}| 1?0 | 2?0 | ... | ...
| ... | 3?1 | ... | salts }| | | | | | | -------------------+--------+-------+------+------+--
----+-----------+--------+
```

```
-------------------+---------+-------+--------+-----------+-------------+  |  |  |  |  |
|Pacquico.|Latham |Saldanha|Australian.|Kooriamooria.| |  |Island.|Bay. | |
```

Water	8·8	24·6	21·3	13·0	8·1
Organic matter and ammoniacal salts	23·0	10·6	14·3	13·7	7·2
Phosphates	32·6	54·7	56·0	44·7	44·5
Sulphate of lime	2·2	2·2	...	4·5	3·9
Carbonate of lime	...	2·0	...	8·2	3·7
Alkaline salts	25·3	4·6	6·0	7·4	11·3
Sand	7·1	0·1	1·4	7·5	21·3
	100·0	100·0	100·0	100·0	100·0
Ammonia	6·8	1·6	1·2	1·1	0·2
Phosphoric acid in alkaline salts	3·0	...	...	...	...

	Patagonian.	Chilian.	Mexican.
Water	20·1	14·9	18·0
Organic matter and ammoniacal salts	19·2	16·1	12·8
Phosphates	30·6	36·0	18·8
Sulphate of lime	1·0	...	27·9
Carbonate of lime	3·6	10·8	...
Alkaline salts	7·1	6·4	16·5
Sand	17·4	14·6	5·0
	100·0	100·0	100·0
Ammonia	2·9	1·2	0·2
Phosphoric acid in alkaline salts	3·0	...	...

Table shewing the Composition of some of the less common varieties of Guano.

NOTE.--The numbers in this Table are mostly derived only from a single analysis and have no value as determining the average composition of these Guanos, but they serve to give a general idea of their value.

	Holme's Island.	Ascension.	Possession Island.	Bear Bay.	Bird Island.	Sea Island.	Indian Bay.
Water	30·2	23·2	25·0	15·7	10·2		
Organic matter and ammoniacal salts	31·8	60·5	32·0	23·5	15·2		
Phosphates	24·3	7·8	27·6	32·4	46·1		
Sulphate of lime	3·4	...	...	...	7·6		
Carbonate of lime	0·8	2·9	...	...	...		
Alkaline salts	7·8	5·8	8·2	15·2	6·5		
Sand	1·7	0·8	6·2	12·2	13·4		
	100·0	100·0	100·0	100·0	100·0		
Ammonia	10·5	10·7	7·5	6·6	1·4		
Phosphoric							

	Sierra Leone	Algoa Bay	New Island	Bird's Island
Water	30?5	28?8	16?2	23?5
Organic matter and ammoniacal salts	6?5	13?8	14?4	4?7
Phosphates	21?4	22?6	25?1	13?8
Sulphate of lime	36?2	...	40?7	29?5
Carbonate of lime	...	13?8	...	...
Alkaline salts	3?2	12?2	1?6	5?0
Sand	1?2	11?8	1?0	23?5
	100?0	100?0	100?0	100?0
Ammonia	0?4	0?4	1?6	0?7
Phosphoric acid in alkaline salts	...	...	1?2	...

On examining the tables given above, it is obvious that guanos may be divided into two classes, the one characterized by the abundance of ammonia, the other by that of phosphates; and which, for convenience sake, may be called ammoniacal and phosphatic guanos. Peruvian and Angamos are characteristic of the former, and Saldanha Bay and Bolivian of the latter class. The value of these two classes of guano differs materially, and they are also applicable under different circumstances, but to these points reference will afterwards be made.

Very special precautions are necessary on the part of the farmer in order to insure his obtaining a guano which is not adulterated, and of good quality if genuine. In the case of Peruvian guano, which is tolerably uniform in its qualities, it is possible to form some opinion by careful examination, and the following points ought to be attended to:

1st, The guano should be light coloured. If it is dark, the chances are that it has been damaged by water.

2d, It should be dry, and when a handful is well squeezed together it should cohere very slightly.

3d, It should not have too powerful an ammoniacal odour.

4th, It should contain lumps, which, when broken, appear of a paler colour than the powdery part of the sample.

5th, When rubbed between the fingers it should not be gritty.

6th, A bushel of the guano should not weigh more than from 56 to 60 lbs.

These characters must not, however, be too implicitly relied on, for they are all imitated with wonderful ingenuity by the skilful adulterator, and they are applicable only to Peruvian guano; the others being so variable that no general rules can be given for determining whether they are genuine. Neither are they so precise as to enable us to give any opinion regarding the relative values of several samples where all are genuine. The only way in which adulteration can with certainty be detected, and the value of different guanos be determined, is by analysis, and the importance of this can easily be illustrated.

In the table above, the average composition of the different guanos is given; but in order to shew how much individual cargos may differ from the mean, we give here analyses of samples of the highest and lowest quality of the genuine guanos of most importance:

	Peruvian.		Bolivian.		Angamos.	
	Highest.	Lowest.	Highest.	Lowest.	Highest.	Lowest.
Water	12?0	7?9	10?7	21?9	11?3	16?0
Organic matter and ammoniacal salts	65?2	50?3	55?3	46?6	11?7	12?6
Phosphates	10?3	8?0	25?0	18?3	62?9	52?5
Alkaline salts	7?0	16?0	7?0	10?4	9?3	13?3
Sand	3?5	17?8	1?0	2?8	4?8	4?6
	100?0	100?0	100?0	100?0	100?0	100?0
Ammonia	25?3	17?5	18?5	14?5	1?9	2?3

The differences are here exceedingly large; and when the values of the two Peruvian guanos are calculated according to the method to be afterwards described, it appears that the highest exceeds the lowest in value by nearly ? per ton. Of course, this is an extreme case, but it is no uncommon occurrence to find a difference of ? or even ? per ton between the values of cargos of

Peruvian guano, which are sold at the same price.

The adulteration of guano is carried on to a very large extent; and though perhaps not quite so extensively now as it was some years since, it is only kept in check by the utmost vigilance on the part of the purchaser. The chief adulterations are a sort of yellow loam very similar in appearance to guano, sand, gypsum, common salt, and occasionally also ground coprolites and inferior guano. These substances are rarely used singly, but are commonly mixed in such proportions as most closely to imitate the colour and general appearance of the genuine article. The extent to which the adulteration is carried may be judged of from the following analyses taken at random from those of a large number of guanos, all of which were sold as first-class Peruvian.

Water 12?5 15?9 12?6 27?6 6?2 Organic matter and } ammoniacal salts } 26?4 44?1 34?4 30?1 27?2 Phosphates 15?4 20?5 22?8 22?7 33?1 Sulphate of lime ... ... 11?8 ... 22?1 Alkaline salts 6?7 9?0 12?1 7?2 22?0 Sand 38?0 10?5 7?3 1?4 10?5 ------ ------ ------ ------ ------ 100?0 100?0 100?0 100?0 100?0

Ammonia 9?4 13?0 9?7 8?4 9?6

In all those cases a very large depreciation in the value has taken place, and several of them are worth considerably less than half the price of the genuine guano, while they are generally offered for sale at about ? under the usual price. The adulteration is chiefly practised in London, and cases occasionally occur which can be traced to Liverpool and other places; but it always takes place in the large towns, because it is only there that facilities exist for obtaining the necessary materials and carrying it out without exciting suspicion. The sophisticated article then passes into the hands of the small country dealers, to whom it is sold with the assurance that it is genuine, and analysis quite unnecessary. In other instances, adulterated and inferior guanos are sold by the analysis of a genuine sample, and sometimes an analysis is made to do duty for many successive cargos of a guano which, though all obtained from one deposit, may differ excessively in composition. In order to insure obtaining a genuine guano, it is above all things important to deal only with a person of established character, who will generally, for his own sake, satisfy himself that the article he vends is genuine and of good quality; and it is always important that the buyer should examine the analysis, and in all cases where there is the slightest doubt, should ascertain that the bulk sent corresponds with it. In the case of a Peruvian guano, a complete analysis is not necessary for this purpose; but an experienced chemist, by the application of a few tests, can readily ascertain whether the sample is genuine. Where the difference in value between different samples is required, a complete analysis is necessary, and this is indispensable in the case of the inferior guanos. Many of these are obtained from deposits of limited extent, and in loading it considerable quantities of the subjacent soil are taken up, so that very great differences may exist even in different parts of the same cargo. Nor must it be forgotten that, except in the case of Peruvian, the name is no guarantee for the quality of the guano, even if genuine. Peruvian guano is all obtained from the same deposits, those of the Chincha Islands, but the guanos which are brought into the market under the name of Patagonian, Chilian, etc., are obtained from a great variety of deposits scattered along the coasts of these countries, sometimes at a distance of several hundred miles from each other, and which have been accumulated under totally different circumstances. In illustration of this, it is only necessary to refer to the subjoined analysis of samples, all of which I believe to be genuine as imported, and which were sold under the name of Upper Peruvian Guano.

I. II. III. Water 7·0 6·5 8·5 Organic matter and ammoniacal salts 10·5 19·6 10·0 Phosphates 67·0 20·1 17·0 Carbonate of lime ... 21·5 ... Alkaline salts 11·0 5·1 61·0 Sand 3·5 27·2 2·5 ------ ------ ------ 100·0 100·0 100·0

Ammonia 2·9 5·3 1·8 Phosphoric acid in the alkaline salts 2·4 ... 1·0 Equal to phosphate of lime 4·9 ... 3·0

With the exception of Peruvian, the supply of good guanos of uniform composition is by no means large, and phosphatic guanos of good quality are now especially rare. The Saldanha Bay, and other similar deposits, have been exhausted, and few guanos of equally good quality have been lately discovered. There is no doubt, however, that such guanos are very useful, and if obtained in large quantity, and of uniform composition, would be used to a much larger extent than they at present are.

The value and use of guano are now so well understood, that it is scarcely necessary to enlarge on the mode of its application. Peruvian guano owes its chief value to its ammonia and phosphates, but it also contains potash, soda, and all the other constituents of plants in small quantity, although in a readily available condition, as is seen in the detailed analysis given in page 205.

In other guanos which have undergone more complete decomposition, and from which the soluble matters have been more or less completely exhausted by rain, the alkaline salts, or at least the potash they originally contained, have almost entirely disappeared. Hence an important difference between Peruvian guano and most other varieties. The former can be used as a complete substitute for farm-yard manure, and excellent crops of turnips and potatoes can be raised by means of it alone, and at a less cost than with ordinary dung. But though this may be done, and in many cases is attended with great economic advantages, it is a practice that cannot be recommended for general use, because the quantity of valuable matters contained in the usual application of guano is much smaller than in farm-yard manure, and the probability is that it would not, if used alone during a succession of years, be sufficient to maintain the soil permanently in a high state of fertility. Five cwt. of Peruvian guano, which is a liberal application per acre, contains about 95 lbs. of ammonia, and 130 of phosphates, while 20 tons of good farm-yard manure contain 312 of ammonia, and about the same quantity of phosphates, and when the other constituents, such as potash and soda, are compared with those in guano, the difference is still more striking. On the other hand, guano is a rapidly acting manure; its constituents are in a condition in which they are more immediately accessible by the plant, and its immediate effect is far more marked, as it is chiefly expended on the crop to which it is applied. It has indeed been alleged that it produces no effects on the subsequent crops, but this opinion can scarcely be considered as well founded. In no case does the crop raised by means of it contain the whole of the ammonia or phosphates present in the manure, and the unappropriated quantity, though it may, and probably does, escape from the lighter soils, must be retained and preserved for the use of subsequent crops by heavy and retentive clay soils. The general inference is, that though guano may at an emergency be used as an entire substitute for farm-yard manure, the practice is one to be generally avoided. When, however, as occasionally happens after a long continued use of farm-yard manure, organic matters have accumulated in the soil, and passed into an inert condition, then Peruvian guano may be used alone with very great advantage. In all cases the rapidity of the action of guano makes it an important auxiliary of farm-yard manure, and it is in this way that it may be most advantageously employed. Experience has shewn that one-half the farm-yard manure may be replaced by guano with the production of a larger crop than by the former alone in its full quantity. The proportion of guano usually employed is from three to five cwt., and it is alleged that a much larger quantity produces prejudicial effects on the subsequent crops, although it is not very easy to see on what this depends.

The variety of guano to be selected must depend to a great extent on the use to which it is to be put. Peruvian guano is most advantageously applied as a top-dressing to young corn and particularly to oats. For the turnip, the ammoniacal guanos were formerly preferred, and on strong soils, under good cultivation, their effects are excellent, but on light soils they are less applicable, their soluble salts being more rapidly washed out, and their effects lost, and in these cases they are surpassed by the phosphatic guanos.

No definite rules can be given for determining the soils on which these different varieties are most applicable, but each individual must determine by experiment that which best suits his own farm; and the inquiry is of much importance to him, as, of course, if the phosphatic guanos will answer as well as the ammoniacal, there is a large saving in the cost of the manure. A very excellent practice is to employ a mixture of equal parts of the two sorts of guano.

Pigeons' Dung.--The dung of all birds, which more or less closely resembles guano, may be employed with much advantage as a manure, but that of the pigeon and the common fowl are the only ones which can be got in quantity. Pigeons' dung, according to Boussingault, contains 8? per cent of nitrogen, equivalent to 10? of ammonia. Its value, therefore, will be more than half that of guano, but it varies greatly, and a sample imported from Egypt into this country, and analysed by Professor Johnston, contained only 5? per cent of ammonia. Hens' dung has not been accurately analysed, but its value must be about the same as pigeons'.

Urate and Sulphated Urine.--We have already discussed the urine of animals, in reference to farm-yard manure. But human urine, the composition of which was then stated, is of much higher value than that of the lower animals, and many attempts have been made to preserve and convert it into a dry manure. Urate is prepared by adding gypsum to urine, and collecting and drying the precipitate produced. It contains a considerable quantity of the phosphoric acid of the urine, but very little of its ammonia; and as the principal value of urine depends on the latter, it is necessarily a very inefficient method of turning it to account. A better method has been proposed by Dr. Stenhouse, who adds lime-water to the urine, and collects the precipitate, which, when dried in the air, contains 1?1 per cent of nitrogen, and about 41 per cent of phosphates. This method is subject to the same objection as that by which urate is made, namely, that the greater part of the ammonia is not precipitated. This might probably be got over to some extent by the addition of sulphate of magnesia, or, still better, of chloride of magnesium, which would throw down the phosphate of magnesia and ammonia. By much the best mode of employing urine is in the form of sulphated urine, which is made by adding to it a sufficient quantity of sulphuric acid to neutralize its ammonia, and evaporating to dryness. In this form all the valuable constituents are retained, and excellent results are obtained from it. Its effects, though mainly attributable to its ammonia, are also in part dependent on the phosphates and alkaline salts which it contains; and it is therefore capable of supplying to the plant a larger number of its constituents than the animal matters already mentioned.

Night-Soil and Poudrette.--The value of night-soil, which is well known, depends partly on the urine, and partly on the f鎐es of which it is formed. Its disagreeable odour has prevented its general use, and various methods have been contrived both for deodorising and converting it into a solid and portable form. The same difficulties which beset the conversion of urine into the solid form occur here, and in most of the methods employed the loss of ammonia is great. It is sometimes mixed with lime or gypsum, and dried with heat, and sometimes with animal charcoal or peat charcoal. The manufacture of a manure from night-soil, called "poudrette," has long been practised in the neighbourhood of Paris and other continental towns. The process employed at Montfau鐇n and at Bondy is very simple. The contents of the cesspools are conveyed to the work in large barrels, which are then emptied into tanks capable of containing the accumulation of several months. When filled they are allowed to stand for some time, during which the smell diminishes and the contents become nearly dry. The residue is then dug out and mixed with ashes, dry loam, charcoal powder, peat, peat-charcoal, saw-dust, and other matters, so as to deodorize it, and render it sufficiently dry for transport. Its general composition may be judged of from the subjoined analyses of samples from different places:--

Montfau鐵 n. Bondy. Dresden. American. Water 28?0 13?0 19?0 39?7 Organic matters 29?0 24?0 20?0 20?7 Phosphates 7?5 4?6 5?0 1?8 Carbonates of lime and } Magnesia, alkaline } 7?5 14?4 11?0 7?3 salts, etc. } Sand 28?0 43?0 43?0 29?5 ------ ------ ------ ------ 100?0 100?0 100?0 100?0

Ammonia 1?4 1?8 2?0 1?3

These analyses shew sufficiently the extent to which the animal matters have been mixed with valueless driers, the second and third samples containing considerably more than half their weight of worthless matters.

Hair, Skin, and Horn.--The refuse of manufactories in which these substances are employed, are frequently used as manures. They are highly nitrogenous substances, and owe their entire value to the nitrogen they contain, their inorganic constituents being in too small quantity to be of any importance, wool and hair having only 2 per cent, and horn 0? per cent of ash. In the pure and dry state, and after subtraction of the ash, their composition is,--

Skin. Human hair. Wool. Horn. Carbon 50?9 50?5 50?5 51?9 Hydrogen 7?7 6?6 7?3 6?2 Nitrogen 18?2 17?4 17?1 17?8 Oxygen 23?2 20?5 } 24?1 24?1 Sulphur ... 5?0 } ------ ------ ------ ------ 100?0 100?0 100?0 100?0

It rarely if ever happens, however, that the refuse offered for sale as a manure is pure. It always contains water, sand, and other foreign matters. Woollen rags are mixed with cotton which has no manurial value, and the skin refuse from tan-works contains much lime. Due allowance must therefore be made for such impurities which are sometimes present in very large quantity.

Refuse horse hair generally contains 11 or 12 per cent of nitrogen. Woollen rags of good quality contain 12? per cent of nitrogen; woollen cuttings about 14; and what is called shoddy only 5? per cent. Horn shavings are extremely variable in their amount of nitrogen; when pure, they sometimes contain as much as 12? per cent, but a great deal of the horn shavings from comb manufactories, etc., contain much sand and bone dust, by which their percentage of nitrogen is greatly diminished, and it sometimes does not exceed 5 or 6 per cent.

All these substances are highly valuable as manures, but it must be borne in mind that they undergo decomposition very slowly in the soil, and hence are chiefly applicable to slow growing crops, and to those which require a strong soil. Woollen rags have been largely employed as a manure for hops, and are believed to surpass every other substance for that crop. As a manure applicable to the ordinary purposes of the farm they have scarcely met with that attention which they deserve, probably because their first action is slow and the farmer is more accustomed to look to immediate than to future results; but they possess the important qualification of adding permanently to the fertility of the soil.

Blood is a most valuable manure, but it is not much employed in this country, at least in the neighbourhood of large towns, as there is a demand for it for other purposes, and it can rarely be obtained by the farmer in large quantity, and at a sufficiently low price. In its natural state it contains about 3 per cent of nitrogen, and after being dried up, the residue contains about 15 per cent. It is best used in the form of a compost with peat or mould, and this forms an excellent manure for turnips, and is also advantageously applied as a top-dressing to wheat.

Flesh.--The flesh of all animals is useful as a manure, and is especially distinguished by the rapidity with which it undergoes decomposition, and yields up its valuable matters to the plant. It is rarely employed in its natural state, but horse flesh was at one time converted into a dry and portable manure, although, I understand, this manufacture is not now prosecuted. The dead animal after being skinned is cut up and boiled in large cauldrons until the flesh is separated from the bones. The latter are removed, and the flesh dried upon a flat stove. The flesh as sold has the following composition:--

Water 12?7 Organic matter 78?4 Phosphate of lime, etc. 3?2 Alkaline salts 3?4 Sand 1?3 ------ 100?0 Nitrogen 9?2 Ammonia to which the nitrogen is } equivalent } 11?0

The dried flesh and small bones of cattle, from the great slaughtering establishments of South America, was at one time imported into this country under the name of flesh manure. Its composition was--

Water 9·5 Fat 11·3 Animal matter 39·2 Phosphate of lime 28·4 Carbonate of lime 3·1 Alkaline salts 0·7 Sand 7·8 ------ 100·0 Nitrogen 5·6 Ammonia to which the nitrogen is } equivalent } 6·7

But owing to the large proportion of phosphates contained in it, it may be most fairly compared with bones. It is not now imported, the results obtained from its use being said not to have proved satisfactory, although this statement appears very paradoxical.

Fish have been employed in considerable quantity as a manure. That most extensively employed in this country is the sprat, which is occasionally caught in enormous quantities on the Norfolk coast, and used as an application for turnips. They are sold at 8d. per bushel, and their composition is--

Water 64· Organic matter 33· Ash 2· ----- 100· Nitrogen 1·0 Phosphoric acid 0·1

The refuse of herring and other fish-curing establishments, whales' blubber, and similar fish refuse, are all useful as manure, and are employed whenever they can be obtained. They are not usually employed alone, but are more advantageously made into composts with their own weight of soil, and allowed to ferment thoroughly before being applied.

Many attempts have been made to convert the offal of the great fish-curing establishments, and the inedible fish, of which large quantities are often caught, into a dry manure, which has received the name of "fish guano." The processes employed have consisted in boiling with sulphuric acid and other agents, and then evaporating, or sometimes by simply drying up the refuse by steam heat. A manure made in this way proved to have the following composition:--

Water 8·0 Fatty matters 7·0 Nitrogeneous organic matters 71·6 Phosphate of lime 8·0 Alkaline salts 3·0 Sand 0·4 ------ 100·0

Nitrogen 11·5 Equal to ammonia 13·8 Phosphoric acid in the alkaline salts, } 0·5 equal to 1·1 phosphate of lime }

The expense of manufacturing manures of this description has hitherto acted as a barrier to their introduction. In this country several manufactories have been established, but either owing to this cause, or to the difficulty of obtaining sufficiently large and uniform supplies of the raw material, some of them have not proved successful, but a manufactory is now in operation in Norway, which exports the manure to Germany. It is probable that most of the processes used in this country failed because they were too costly, and it is much to be desired that the subject should be actively taken up. It is said that the refuse from the Newfoundland fisheries is capable of yielding about 10,000 tons of fish guano annually; and the quantity obtainable on our own coasts is also very considerable.

Bones.--Bones have been used as a manure for a long period, but they first attracted the particular attention of agriculturists from the remarkable effects produced by their application on the exhausted pasture lands of Cheshire. During the present century they came into general use on arable land, and especially as a manure for turnips; and they are now imported in large quantities from the continent of Europe. The bones used in agriculture are chiefly those of cattle, but sheep and horse bones are also employed. They do not differ much in quality when genuine. The subjoined analysis is that of a good sample.

Water 6?0 Organic matter 39?3 Phosphate of lime 48?5 Lime 2?7 Magnesia 0?0 Sulphuric acid 2?5 Silica 0?0 ------ 100?0 Ammonia which the organic matter } is capable of yielding } 4?0

In general, bones may be said to contain about half their weight of phosphate of lime, and 10 or 12 per cent of water. But, in addition to their natural state, they are met with in other forms in commerce, in which their organic matter has been extracted either by boiling or burning. The latter is especially common in the form of the spent animal charcoal of the sugar refiners, which usually contains from 70 to 80 per cent of phosphate of lime, but when deprived of their organic matter, they may be more correctly considered under the head of mineral manures.

From the analysis given above, it is obvious that the manurial value of bones is dependent partly on their phosphates and partly on the ammonia they yield. It has been common to attribute their entire effects to the former, but this is manifestly erroneous; and although there are no doubt cases in which the former act most powerfully, the benefit derived from the ammonia yielded by the organic matter is unequivocal. When the phosphates only are of use, burnt bones or the spent animal charcoal of the sugar refiners are to be preferred.

At their first introduction, bones were applied in large fragments, and in quantities of from 20 to 30 cwt., or even more, per acre, but as their use became more general they were gradually employed in smaller pieces, until at last they were reduced to dust, and it was found that, in a fine state of division, a few hundredweights produced as great an effect as the larger quantity of the unground bones. Even the most complete grinding which can be attained, however, leaves the bones in a much less minute state of division than guano, and they necessarily act more slowly than it does, the more especially as they contain no ready-formed ammonia. They may be still further reduced by fermentation, which acts by decomposing the organic matter, and causing the production of ammonia; but not as is frequently, though erroneously supposed, by converting the phosphates into a soluble condition, for this does not occur to any extent, and their more rapid action is solely due to the partial decomposition of the organic matter, by which it is brought into a condition capable of undergoing a more rapid change in the soil. The rapidity of action of bones is still more promoted by solution in sulphuric acid, by which they are converted into the form of dissolved bones or superphosphate. At the present moment, however, very little of the superphosphates sold in the market are made exclusively from bones in their natural state, by far the larger portion being manufactured from mineral phosphates, or from bones after destruction of their organic matter, sometimes with the addition of small quantities of unburnt bones, but more frequently of sulphate of ammonia, to yield the requisite quantity of ammonia. These substances may therefore be best considered under the head of mineral manures.

CHAPTER XI.

COMPOSITION AND PROPERTIES OF MINERAL MANURES.

Mineral manure is a term which is now used with great laxity. In its strict sense, it means manures which contain only, and owe their exclusive value to the presence of, those substances which go to make up the inorganic part or ash of plants. It has, however, been usually taken to include all saline matters, and especially the compounds of ammonia and nitric acid, which are indebted for their manurial effects to the nitrogen they contain; and thus is so far incorrect. It would, however, be manifestly impossible to arrange these compounds with any degree of accuracy among either animal or vegetable manures, and hence the necessity of including them amongst those which are strictly mineral. The most important practical distinction between them and the substances discussed in the two preceding chapters is, that the latter generally contain the whole or the greater part of the constituents of plants. Even bones yield a certain quantity of alkalies, magnesia, sulphuric acid, and chlorine, and may in some sense be considered as a general manure. But those to which the term mineral manure is applied for the most part contain only one or two of the essential elements of plants, and hence cannot be applied as substitutes for the substances already discussed, although they are frequently most important additions to them.

Sulphate and Muriate of Ammonia.--These and other salts of ammonia have been tried experimentally as manures, and it has been ascertained that they may all be used with equal success; but as the sulphate is by much cheaper, it is that which probably will always be employed to the exclusion of every other. It contains, when pure, 25? per cent ammonia.

It is now manufactured of excellent quality for agricultural use, and when good, contains from 95 to 97 per cent of actual sulphate, the remainder consisting chiefly of moisture and a small quantity of fixed residue; but specimens are occasionally met with containing as much as 10 per cent of impurities, which, as its price is high, makes a material difference in its value. Inferior descriptions are also occasionally sold, among which is a variety distinguished by containing a large quantity of water and fixed salts, although it appears to the eye a good article. Its composition is--

I. II. Water 9?5 5?7 Sulphate of ammonia 79?3 85?1 Fixed salts 11?7 9?2 ---------- 100?0 100?0 Ammonia 20?5 21?4

An article called sulphomuriate of ammonia is also sold for agricultural use. It is obtained as a refuse product in the manufacture of magnesia, and is a mixture of sulphate and muriate of ammonia, with various alkaline salts. It differs somewhat in quality, and is sold by analysis at a price dependent on the ammonia it contains.

	I.	II.
Water	14?9	25?9
Sulphate of ammonia	62?5	47?9
Muriate of ammonia	15?	...
Sulphate of soda	...	9?2
Sulphate of magnesia	...	18?8
Chloride of potassium	4?5	2?4
Chloride of sodium	17?5	0?5
	------	------
	100?0	100?0
Ammonia	16?0	11?8

The quality of sulphate of ammonia may generally be judged of from its dry and uniformly crystalline appearance, and it may be tested by heating a small quantity on a shovel over a clear fire, when it ought to volatilize completely, or leave only a trifling residue. Some care, however, is necessary in applying this test, as in the hands of inexperienced persons it is sometimes fallacious. The salts of ammonia may be applied in the same way as guano; but they are most advantageously employed as a top-dressing, and principally to grass lands. In this way very remarkable effects are produced, and within a week after the application, the difference between the dressed and undressed portions of a field is already conspicuous. Experience has shewn that success is best insured when the salt is applied during or immediately before rain, so that it may be at once incorporated with the soil; as when used in dry weather little or no benefit is derived from it. It seems also to exert a peculiarly beneficial effect upon clover; and hence it ought to be employed only on clover-hay, as where ryegrass or other grasses form the whole of the crop we have better manures.

Ammoniacal Liquor of the Gas-Works, and of the Animal Charcoal Manufacturers.--Both of these are excellent forms in which to apply ammonia, when they can be obtained. The ammoniacal liquor of the gas-works is very variable in quality, but contains generally from 4 to 8 ounces of dry ammonia per gallon, which corresponds in round numbers to from 1 to 2 lb. of sulphate of ammonia. It is best applied with the watering-cart, but must be diluted before use with three or four times its bulk of water, as if concentrated it burns up the grass, and it is also advisable to use it during wet weather. The ammoniacal liquor of the ivory-black works contains about 12 per cent of ammonia, or about four or five times as much as gas liquor. It has been used in some parts of England, made into a compost, and applied to the turnip and other crops, and, it is said, with good effect. Bone oil, which distils over along with it, has also been used in the form of a compost; it contains a large quantity of ammonia and of nitrogen in other forms of combination; the total quantity of nitrogen it contains being 9?4 per cent, which is equivalent to 10?8 of ammonia. Only part of this nitrogen is actually in the state of ammonia; and some circumstances connected with the chemical relations of the other nitrogenous compounds in this substance render it probable that they may pass very slowly into ammonia, and may therefore be of inferior value; but the substance deserves a trial, as it is very cheap. It must be carefully composted with peat, and turned over several times before being used.

Nitrates of Potash and Soda.--Nitrate of potash has long been used as a manure, but its high price has prevented its general application, and its place has now been almost entirely taken by nitrate of soda, which is much cheaper and contains weight for weight a larger quantity of nitrogen. Both these salts are employed as sources of nitrogen; but nitrate of potash owes also a certain proportion of its value to the potash it contains. Nitrate of soda, on the other hand, must be considered to owe its entire value to its nitric acid, as soda is of little value to the plant; and, moreover, can be obtained in common salt at a price so low, as to make it a matter of no moment in the valuation of the nitrate. In its ordinary state, as imported from Peru, nitrate of soda contains from 5 to 10 per cent of impurities, and it bears a price proportionate to the quantity of the pure salt present in it. When of good quality it contains about 15 per cent of nitrogen, equivalent to 18 of ammonia, and is, therefore, richer in that constituent of plants than Peruvian guano. It is essentially a rapidly acting manure, and produces a marked effect within a very few days after its application; but owing to the fact that nitric acid cannot be absorbed and retained by the soil in the same manner as ammonia, it is liable to be lost unless it can be at once assimilated by the plant. For this reason it acts best when applied in small quantity as a top-dressing to grass-land, and to young corn. A large application has no advantages, and there can be no doubt that the best effect would be produced by several very small quantities, applied at intervals. In one experiment, Mr. Pusey found 42 lb. per acre to increase the produce of barley by 7 bushels, and very favourable results have been obtained by other experimenters. The beneficial effects of nitrate of soda appear to be almost entirely confined to the grasses and cereals. At least experience here has shewn that it produces little or no effect on clover; and one farmer has stated, that having recently adopted the practice of sowing clover with a very small proportion of ryegrass only, he has been led to abandon the use of nitrate of soda, which he formerly employed abundantly, when ryegrass formed a principal part of his crop. The action of nitrate of soda is very remarkable, not only in this respect, but also because a given quantity of nitrogen in it appears to produce a greater effect than the same quantity in sulphate of ammonia or guano. At the same time this statement must be taken as very general, definite experiments being still too few to admit of its being stated as an absolute fact. The probability is, that the same quantity of nitrogen, in the form either of ammonia or nitrate of soda, will produce the same effect, although the conditions necessary for its successful action may not be the same with the two manures. It is alleged that nitrate of soda is advantageously conjoined with common salt, which is said to check its tendency to make the grain crops run to straw, and to prevent their lodging, as they are apt to do, when it is employed alone. But considerable difference of opinion exists in this point, many farmers believing that salt produces no effect. When employed for hay, especially when mixed with clover, it is advisable to use it along with an equal quantity of sulphate of ammonia, which gives a better result than either separately.

Salts of Potash and Soda.--The substances just mentioned must be considered to owe their chief manurial value to nitric acid; but other salts have been used as manures in which the effect is undoubtedly due to the alkalies themselves. With the exception of common salt, most of the alkaline salts have only been used to a limited extent; and it is remarkable that, so far as our present experience goes, there is no class of substances from which more uncertain results are obtained.

Muriate and Sulphate of Potash have both been used, and the former has in some cases, and in particular seasons, produced a very remarkable effect in the potato; but in other instances it has proved quite useless. The cause of this difference has not been ascertained. Sulphate of soda has also been used to some extent, but apparently without much benefit; and there is no reason to expect that it should act better than common salt, which can be obtained at a much lower price.

Chloride of Sodium, or Common Salt, has at different times been employed as a manure, but its effects are so variable and uncertain, that its use, in place of increasing, has of late years rather diminished, it having frequently been found that on soils in all respects similar, or even on the same soil, in different years, it sometimes proves advantageous, at others positively injurious. Its use as an addition to nitrate of soda has been already alluded to, and it is said that it produces the same effect when mixed with guano and salts of ammonia. The accuracy of this statement is doubted by many persons, and the explanation which has been given of the cause of its action is more than dubious. It is supposed to enable the plant to absorb more silica from the soil; but this is a speculative explanation of its action, and has not been supported by definite experiment. Although little effect has been observed from salt, it deserves a more accurate investigation, as not withstanding the extent to which it has been employed, we are singularly deficient in definite experiments with it.

Carbonates of Potash and Soda have only been tried experimentally, and that to a small extent, nor is it likely that they will ever come into use, owing to their high price. The remarks we have made in the section on the ashes of plants regarding the subordinate value of soda, will enable the reader to see that greater effects are to be anticipated from the former than from the latter of these salts. They may, however, exert a chemical action on the soil, altogether independent of their absorption by the plant, but its nature and amount are still to determine.

Silicates of Potash and Soda have been employed with the view of supplying silica to the plant, but the results have been far from satisfactory. This may perhaps have been due to the doubtful nature of the commercial article, but now that silicate of soda can be obtained of good quality, it is desirable that the experiments should be repeated. It is said to have produced good effects on the potato.

Sulphate of Magnesia can be obtained at a low cost, and has been used as a manure in some instances with very marked success. It has been chiefly applied as a top-dressing to clover hay, but it seems probable that it might prove a useful application to the cereals, the ash of which is peculiarly rich in magnesia.

Many other saline substances have been tried as manures; but in most instances on too limited a scale to permit any definite conclusions as to their value. The experiments have also been too frequently performed without the precautions necessary to exclude fallacy, so that the results already arrived at must not be accepted as established facts, but rather as indications of the direction in which further investigation would be valuable. There is little doubt that many of these substances might be usefully employed, if the conditions necessary for their successful application were eliminated; and no subject is at present more deserving of elucidation by careful and well-devised field experiments.

Phosphate of Lime.--The use of bones in their natural state as a manure has been already adverted to, and it was stated, that though their value depended mainly on the phosphates, the animal matters and other substances contained in them were not without effect. The action of phosphates is greatly promoted by solution in sulphuric acid, and the application of the acid has brought into use many varieties of phosphates of purely mineral origin, or which have been deprived of their organic matters by artificial processes. Of these, the spent animal charcoal of the sugar-refiners, usually containing about 70 per cent of phosphates, and South American bone ash, are the most important. The latter is now imported in very large quantity, and has the composition shewn in the following analyses:--

I. II. III.

Water 6?0 6?8 3?3 Charcoal 5?5 2?9 2?2 Phosphates 79?0 71?0 88?5 Carbonate of lime 4?5 3?5 5?0 Alkaline salts 0?5 traces ... Sand 5?5 16?0 0?0 - ----- ------ ------ 100?0 100?0 100?0

Bone ash has hitherto been almost entirely consumed as a raw material for the manufacture of superphosphates; but as it is sold at from ?: 10s. to ?: 10s. per ton when containing 70 per cent of phosphates, it is, in reality, a very cheap source of these substances, and merits the attention of the farmer as an application in its ordinary state.

Of strictly mineral phosphates, a considerable variety is now in use, but they are employed exclusively in the manufacture of superphosphates, as in their natural state they are so hard and insoluble, that the plant is incapable of availing itself of them.

Coprolites.--This name was originally applied by Dr. Buckland to substances found in many geological strata, and which he believed to be the dung of fossil animals. It has since been given to phosphatic concretions found chiefly in the greensand in Suffolk and Cambridgeshire, which are certainly not the same as those described by Dr. Buckland, but consist of fragments of bones, ammonites, and other fossils. Coprolites are now collected in very large quantities, and about 43,000 tons are annually employed. They are extremely hard, and require powerful machinery to reduce them to powder, and hence their price is considerable, being about ?: 10s. per ton. Their composition varies somewhat according to the care taken in selecting them, and the locality from which they have been obtained. A general idea of their composition may be derived from the subjoined analyses:--

Water 1?5 1?0 Organic matter 2?9 6?5 Phosphate of lime 55?1} 61?5 Phosphate of iron 3?4} Carbonate of lime 26?0 16?0 Sulphate of lime 1?7 " Alkaline salts 1?5 3?1 Sand 5?9 11?5 ------ ------ 100?0 100?0

Within the last two or three years, coprolites have been found in great abundance in France, but they are of inferior quality, and rarely contain more than 40 per cent of phosphates.

Apatite, or mineral phosphate of lime, is found in large deposits in different places. It is particularly abundant in Spain, and occurs also in America and Norway. From the latter country it has been imported to some extent; and during the last year considerable quantities have been brought from Spain, and the importations will undoubtedly increase very largely as the means of transport improve in that country. Spanish apatite contains--

Water 0?0 Phosphate of lime 93?0 Carbonate of lime 0?0 Chlorine, etc. traces Sand 4?0 ----- 99?0

Several other varieties of mineral phosphates have been imported under the name of guano. The most important is Sombrero Island guano, which is found on a small island in the Gulf of Mexico, where it occurs in a layer said to be forty feet thick. It contains--

Water 8?6 Phosphate of lime 37?1 Phosphates of alumina and iron 44?1 Phosphate of magnesia 4?0 Sulphate of lime 0?6 Carbonate of lime 3?6 Sand 0?0 ------ 100?0

A somewhat similar substance, but in hard crusts, has been imported, under the names of Maracaybo guano, Pyroguanite, etc., which contains--

Water 1·3 Organic matter 6·8 Phosphates 75·9 Alkaline salts 4·1 Sand 11·4 ------ 100·0 Phosphoric acid in the alkaline } 0·8 salts = 1·8 phosphate of lime }

These substances are all excellent sources of phosphates, but they are so hard that the plants cannot extract phosphoric acid from them, and they are only useful when made soluble by chemical processes.

Superphosphate; Dissolved Bones.--These names were at first applied to bones which had been treated with sulphuric acid; but superphosphates are now rarely made from bones alone, but bone ash and some of the mineral phosphates just described are employed, either along with them, or very frequently alone. The manufacture of superphosphates depends on the existence of two different compounds of phosphoric acid and lime, one of which contains three times as much lime as the other. That which contains the larger quantity of lime is found in the bones and all other natural phosphates, and is quite insoluble in water; but when two-thirds of its lime are removed, it is converted into the other compound, which is exceedingly soluble. This change is effected by the use of sulphuric acid, which combines with two-thirds of the lime of the ordinary insoluble phosphate of lime, and converts it into the biphosphate of lime, which is soluble. When, therefore, we add to 100 lbs. of common phosphate of lime the necessary quantity of sulphuric acid, it yields 64 lbs. of biphosphate, containing the whole of the phosphoric acid, which is the valuable constituent, the diminution in weight being due to the removal of the valueless lime. Hence it follows, also, that as the lime so removed is converted into sulphate, there must, for every 100 lbs. of phosphate of lime converted into biphosphate, be produced 87 lbs. of dry sulphate of lime, or 110 of the ordinary sulphate called gypsum. This is the minimum quantity which can be present, but in actual practice it is liable to be greatly exceeded, more especially where coprolites are used, owing to the large amount of carbonate of lime they contain, which is also converted into sulphate by the action of the acid, so that it is far from uncommon to find the gypsum twice as great as it would be if materials free from carbonates could be obtained. By employing a sufficiency of sulphuric acid, the whole quantity of phosphoric acid in the bones may be thus brought into a soluble state, but in actual practice it is found preferable to leave part of it in the insoluble condition; as where it is entirely soluble, its effect is too great during the early part of the season, and deficient at its end. In order to dissolve bones, bone ash, or mineral phosphates, they are mixed with from a third to half their weight of sulphuric acid, of specific gravity 1?0 or 140?Twaddell. When mineral phosphates, and particularly coprolites, are used, the quantity of sulphuric acid must be increased so as to compensate for the loss of that which is consumed in decomposing the carbonate of lime they contain. When operating on the small scale, the materials are put into a vessel of wood, stone, or lead (iron is to be avoided, as it is rapidly corroded by the acid), and mixed with from a sixth to a fourth of their weight of water, which may with advantage be used hot. The sulphuric acid is then added, and mixed as uniformly as possible with the bones. Considerable effervescence takes place, and the mass becomes extremely hot. At the end of two or three days it is turned over with the spade, and after standing for some days longer, generally becomes pretty dry. Should it still be too moist to be sown, it must be again turned over, and mixed with some dry substance to absorb the moisture. For this purpose everything containing lime or its carbonate must be carefully avoided, as they bring back the phosphates into the insoluble state, and undo what the sulphuric acid has done. Peat, saw-dust, sand, decaying leaves, or similar substances, will answer the purpose, and they should all be made thoroughly dry before being used. An excellent plan is to sift the bones before dissolving, to apply the acid to the coarser part, and afterwards to mix in the fine dust which has passed through the sieve, to dry up the mass; or a small quantity of bone ash, of good quality, or Peruvian guano, may be used. On the large scale, mechanical arrangements are employed for mixing the materials, so as to economise labour, and mineral phosphates, such as apatite, can then be used with advantage. In such cases, blood, sulphate of ammonia, soot, and other refuse matters, are occasionally used to supply the requisite quantity of nitrogenous substances, but large quantities are also made from bone ash, etc., without these additions.

The composition of superphosphates must necessarily vary to a great extent, and depends not only on the materials, but on the proportion of acid used for solution. The following analysis illustrates the composition of good samples made from different substances--

+----------------------------------+------------------+------------------+ | | | | | Bones

	Bone alone.		Bone-Ash.	
Water,	7·4	... 7·9	5·3	... 10·0
Organic matters and ammoniacal salts,	17·3	... 21·9	6·4	... 4·2
Biphosphate of lime	13·8	... 9·7	21·5	... 23·9
Equivalent to soluble phosphates,	(20·7)	...(15·9)	(33·3)	...(36·2)
Insoluble phosphates	10·1	... 21·7	5·2	... 6·8
Sulphate of lime,	46·0	... 35·0	56·6	... 47·8
Alkaline salts,	1·6	... 0·4	trace.	
Sand,	3·8	... 3·0	4·3	... 4·0
	100·0	...100·0	100·0	...100·0
Ammonia,	2·1	... 3·1	0·3	... 0·1

	Chiefly Coprolites.		Mixtures containing Salts of Ammonia, etc.	
Water,	5·0	... 10·7	7·7	... 15·2
Organic matters and ammoniacal salts,	5·0	... 4·3	9·7	... 13·6
Biphosphate of lime	12·4	... 13·5	17·3	... 12·7
Equivalent to soluble phosphates,	(19·0)	...(21·3)	(27·0)	...(19·7)
Insoluble phosphates	16·0	... 0·7	12·0	... 8·0
Sulphate of lime,	52·9	... 62·2	49·7	... 45·4
Alkaline salts,	2·7	... 0·6	0·6	... 1·7
Sand,	6·0	... 8·0	3·0	... 2·4
	100·0	...100·0	100·0	...100·0
Ammonia,	0·1	... 0·7	1·8	... 1·5

Superphosphates made from bones alone are generally distinguished by a large quantity of ammonia, and a rather low per centage of biphosphate of lime. This is owing to the difficulty experienced in making the acid react in a satisfactory manner on bones, the phosphates being protected from its action by the large quantity of animal matter which, when moistened, swells up, fills the pores, and prevents the ready access of the acid to the interior of the fragments. Superphosphates from bone-ash, on the other hand, contain a mere trifle of ammonia, and when well made a very large quantity of biphosphate of lime. Their quality differs very greatly, and depends, of course, on that of the bone-ash employed, which can rarely be obtained of quality sufficient to yield more than 30 or 35 per cent of soluble phosphates. Coprolites are seldom used alone for the manufacture of superphosphates, but are generally mixed with bone-ash and bone dust. Mixtures containing salts of ammonia, flesh, blood, etc., are also largely manufactured, and some are now produced containing as much as four or five per cent of ammonia, and the consumption of such articles is largely increasing.

The analyses above given are all those of good superphosphates, in which abundance of acid has been used so as to convert a large proportion of insoluble into soluble phosphates; but there are many samples of very inferior quality to be met with in the market, in which the proportion of acid has been reduced, and the quantity of phosphates made soluble is consequently much lower than it ought to be. The following analyses illustrate the composition of such manures, which are all very inferior and generally worth much less than the price asked for them.

Water 21·0 5·7 7·9 Organic matter and ammoniacal salts, 11·2 13·1 8·0 Biphosphate of lime 2·8 2·2 6·2 Equivalent to soluble phosphates (4·5) (3·5) (10·2) Insoluble phosphates 25·0 15·0 14·3 Sulphate of lime 23·6 47·2 51·3 Alkaline salts 10·0 3·3 3·3 Sand 3·0 11·5 8·0 ------ ------ ------ 100·0 100·0 100·0 Ammonia, 1·2 0·9 0·3

The deliberate adulteration of superphosphate, that is, the addition to it of sand or similar worthless materials, I believe to be but little practised. The most common fraud consists in selling as pure dissolved bones, articles made in part, and sometimes almost entirely, from coprolites. Occasionally refuse matters are used, but less with the intention of actually diminishing the value of the manure as for the purpose of acting as driers. It is said that sulphate of lime is sometimes employed for this purpose, but this is rarely done, because that substance is always a necessary constituent of superphosphate in very large quantities; and as farmers look upon it with great suspicion, all the efforts of the manufacturers are directed towards reducing its quantity as much as possible. It is very commonly supposed by farmers that the sulphate of lime found in so large quantity in all superphosphates, and often amounting to as much as fifty per cent, has been added to the materials in the process of manufacture, but this is a mistake; it is a necessary and inevitable product of the chemical action by which the phosphates are rendered soluble, although its quantity depends on the materials from which the manure is made. When pure bones are used its quantity is small, and it does not greatly exceed twice that of the biphosphate of lime; but in a manure made from coprolites, or other substances containing a large proportion of carbonate of lime, which must in the process of manufacture be converted into sulphate, it may be four or five times as much.

Although there is no manure which varies more in quality, or requires greater vigilance on the part of the purchaser, in order to obtain a good article, there is no doubt that superphosphates, owing to the process of manufacture being better understood, and to increased competition, have considerably improved in quality. Six or eight years since a manure containing thirty per cent of phosphates, of which twelve or fifteen had been converted into biphosphate, was considered a fair sample, but now the proportion rendered soluble is greatly increased; and where bone ash alone is employed, as much as thirty and even forty per cent of soluble phosphates is occasionally found. This, of course, is an exceptional case, and great attention and care in the selection of materials are necessary to obtain so large a proportion. The analyses already given will shew the farmer what he has to expect in good superphosphates, but it is very necessary that he should take care to obtain from the manufacturer a manure equal to the guarantee; and he ought to bear in mind that, owing to the difficulty of getting materials of constant composition, variations often take place to a considerable extent in manures which are supposed to be made in exactly the same manner.

Phospho-Peruvian Guano.--Under this name a kind of superphosphate, which is understood to be made by dissolving a native "rock guano," has recently attracted considerable attention, and is used to a large extent. Its composition is--

Water 9?4 Organic matter 21?8 Biphosphate of lime, equivalent to 25?2 soluble phosphates 16?1 Insoluble phosphates 10?8 Sulphate of lime 37?1 Alkaline salts, containing 1?2 of phosphoric acid, and equivalent to 2?6 soluble phosphates 2?2 Sand 1?1 ------ 100?0 Ammonia, 3?0

It is chiefly distinguished by the large proportion of valuable ingredients it contains, and the care taken to secure uniformity of composition.

A variety of substances are sold under the name of nitrophosphate, potato manure, cereal manure, etc. etc., which are all superphosphates, differing only in the proportion of their ingredients, and in the addition of small quantities of alkaline salts, sulphate of magnesia, and other substances, but they present little difference from ordinary superphosphates in their effects.

The use of superphosphate has greatly extended of late years, and its consumption has increased in a greatly more rapid ratio than that of guano or any other manure. Ten or twelve years since it was comparatively little known, but it has now come to be used in many cases in which Peruvian guano was formerly employed. It produces a better effect than that manure on light soils, although in general a mixture of the two answers better than either separately. When Peruvian guano is to be applied along with it, the farmer will naturally select a superphosphate made from bone ash, and containing the largest obtainable quantity of soluble phosphates; but when it is to be used alone, it is advisable to take one made from bones, or at all events one containing a considerable quantity of nitrogenous matter or ammonia. The kind to be selected must, however, be greatly dependent on the particular soil, and the situation in which it is to be used.

Lime.--Lime is by far the most important of the mineral manures, and an almost indispensable agent of agricultural improvement. It has been used as chalk, marl, shell and coral sand, ground limestone, and as quick and slaked lime, and its action varies according as it is applied in any of its natural forms, or after being burnt. In all of its native forms the lime is combined with carbonic acid in the proportion of fifty-six parts of lime to forty-four of carbonic acid, and the carbonate is generally mixed with variable quantities of earthy ingredients, which in some instances are important additions to it, and affect its utility as a manure.

Chalk is a very pure form of carbonate of lime, and where it abounds has been largely employed as an application on the soil. It is dug out of pits and exposed to the action of the winter's frost, by which it is thoroughly disintegrated, and in spring it is applied in quantities, which, in many instances, are only limited by the question of cost.

Marl is a name given to a mixture of finely-divided carbonate of lime, with variable proportions of clay and siliceous matters, which is found at the bottom of valleys and in hollow places in beds often of considerable extent and thickness, where it is deposited from the waters of lakes holding lime in solution, fed by streams passing over limestone, or rocks rich in lime. The composition of marls differs greatly in different districts, and they have been divided into true marls, and clay marls, according as the carbonate of lime or clay is the preponderating ingredient. The following table illustrates the composition of different varieties:--

	Luneburg	Ayrshire	Wesermarsh	Barbadoes
Carbonate of lime	93?	85?	8?	8?
Carbonate of magnesia	...	1?	...	3?
Sulphate of lime	...	0?	...	0?
Phosphate of lime	0?	2?	...	1?
Alumina and oxide of iron	1?	4?	2?	7?
Alkaline salts	...	0?	...	1?
Silica and clay	4?	5?	84?	78?
Organic matter	0?	0?	2?	...
Water	...	...	1?	...
	100?0	100?0	99?	100?0

The true marls, that is those in which carbonate of lime abounds, are greatly preferable to clay marls, the latter, indeed, operate chiefly mechanically, by altering the texture of the soil--the lime they contain being frequently too small to exercise much appreciable effect.

Shell and coral sands consist chiefly of fragments of shells and coral disintegrated by the action of the waves, and mixed with more or less siliceous sand, and containing small quantities of phosphate of lime. They occur to a considerable extent both on our own coasts and those of France, and have been used with good effect on some descriptions of soil.

The general composition of limestones has been already adverted to, when treating of the origin of soils, and a distinction drawn between the common limestones and dolomite or magnesium limestone. Few limestones can be considered as even approaching to purity, and they almost all contain a small quantity of carbonate of magnesia as well as earthy matters, and occasionally a little phosphate of lime. In good specimens the quantities of these substances are generally small, and they usually contain about half their weight of lime. When limestone is burnt in the kiln, the change which ensues consists in the expulsion of the carbonic acid, and the consequent conversion of the lime into the uncombined or quick state. If water be thrown upon it when in this condition, it becomes hot, swells up, and falls to a fine soft powder, and has then entered into combination with water. If it be exposed to the air, the same action takes place, although, of course, more slowly; and if it be left for a sufficient time, it at length absorbs carbonic acid, and reverts to its original form of carbonate of lime, although now in a state of very fine division.

While lime may be applied in the state of carbonate, either as chalk, marl, or pounded limestone, and with a certain amount of advantage, much greater effects are obtained from the use of lime itself in the quick or slaked state. These advantages are dependent partly on the mechanical effect of the burning and slaking, which enable us to reduce the lime to a much more minute state of division, and consequently to incorporate it more uniformly and thoroughly with the soil, and partly on the more powerful chemical action which it exists when in the quick or caustic state. Other minor advantages are also secured, such as the production of a certain quantity of sulphate of lime, produced by the oxidation of the sulphur of the coal used in burning, etc., which, though comparatively trifling, may, under particular circumstances and in some soils, be of considerable importance.

The action of lime is of a complicated character. Where the soil is deficient in lime, it must necessarily act by supplying that substance to the plants growing in it. But this is manifestly a very subordinate part of its action,--1st, Because no soil exists which does not contain lime in sufficient quantity to supply that element to the plants. 2d, Because its effects are not restricted to those soils in which it exists naturally in small quantity; and, 3d, Because it is found that a small application, such as would suffice for the wants of the crops, is not sufficient to produce its best effects.

It is a familiar fact that the quantity of lime applied to the soil for agricultural purposes is very large, as much as ten, and even twenty tons per acre having been used, while the smallest application is exceedingly large when compared with the mere requirements of the crops. Of late years the very large applications once in use have become less common, as it has been found preferable to employ smaller doses more frequently repeated. The quantity used depends, however, to a great extent, on the nature and condition of the soil, heavy clays, especially if undrained, and soils of a peaty nature, requiring a large application; while on well drained and light soils a smaller quantity suffices. Thin soils also require only a small application. The geological origin of the soil is also not without its influence, and its beneficial effect is peculiarly seen on granite, porphyry, and gneiss soils, both because these are naturally deficient in lime, and because the decompositions by which their valuable constituents are liberated take place with extreme slowness.

The greater part of the action of lime is unquestionably dependent on its exerting a chemical decomposition on the soil; and it acts equally on both the great divisions of its constituents, the inorganic and the organic. On the former, it operates by decomposing the silicates, which form the main part of the soil, and the alkalies they contain being thus set free, a larger supply becomes available to the plant. On the organic constituents its effects are principally expended in promoting the decomposition which converts their nitrogen into ammonia; and thus a supply of food, which might remain for a long period locked up, is set free in a state in which the plant can at once absorb it. But these chemical decompositions are attended by a corresponding change in the mechanical characters of the soil. Heavy clays are observed to become lighter and more open in their texture; and those which are too rich in organic matter have it rapidly reduced in quantity, and the excessive lightness which it occasions diminished.

The effects of an application of lime are not generally observed immediately, but become apparent in the course of one or two years, when it has had time to exert its chemical influence on the soil; but from that time its effects are seen gradually to diminish and finally to cease entirely. The period within which this occurs necessarily varies with the amount of the application and the nature of the soil, but it may be said generally that lime will last from ten to fifteen years. The cessation of its effects is due to several circumstances, partly of course to the absorption of lime by the plants, partly to its being washed out of the soil by the rains, and partly to its tendency to sink to a lower level, a tendency which most practical men have had opportunities of observing. In the latter case, deep-ploughing often produces a marked effect, and sometimes makes it possible to postpone for a year or two the reapplication of lime. All these circumstances have their influence in bringing its action to an end, but the most important is, that after a time it has exhausted its decomposing effect on the soil, having destroyed all the organic matter, or liberated all the insoluble mineral substances which the quantity added is competent to do, and so the soil passes back to its old state. It does even more, for unless active measures are taken to sustain it by other means, it is found that the fertility of the soil is apt to become less than it was before the use of lime. And that it should be so is manifest, if we consider that the lime added has liberated a quantity of inorganic matter, which, in the natural state of the soil, would have become slowly available to the plant, and that it must have acted chiefly in those very portions which, from having already undergone a partial decomposition, were ready to pass into a state fitted for absorption, and thus as it were must have anticipated the supplies of future years. This effect has been frequently observed by farmers, and is indeed so common, that it has passed into a proverbial saying, that "lime enriches the fathers and impoverishes the sons." But this is true only when the soil is stinted of other manures, for when it is well manured the exhausting effect of lime is not observed; and it must be laid down as a practical rule, that its use necessitates a liberal treatment of the soil in all other respects. But when lime has been once employed it becomes almost necessary to resort to it again; and generally so soon as its effects are exhausted a new quantity is applied, not so large as that which is used when the soil is first limed, but still considerable. When this is done very frequently, however, bad effects ensue; the soil gets into a particular state, in which it is so open that the grain crops become uncertain, and such land is said, in practical language, to be overlimed. The explanation of this state of matters commonly assumed by those unacquainted with chemistry is, that the land has become too full of lime; but a moment's consideration of the very small fraction of the soil which even the largest application of lime forms, will serve to shew that this cannot be the cause. Ten tons of lime per acre amounts to only one per cent of the soil, and as a considerable part of the lime is carried off by drainage in the course of years, it is obvious that even very large and frequently repeated doses are not likely to produce any great accumulation of that substance. In point of fact, analyses of overlimed soils have proved that the lime does not exceed the ordinary quantity found in fertile land. The explanation of the phenomenon is probably to be found in the rapid decomposition of organic matter by the lime, and its escape as carbonic acid, by which the soil is left in that curious porous condition so well known in practice. The cure for overliming is found to be the employment of such means as consolidate the soil, such as eating off with sheep, rolling, or laying down to permanent pasture.

The immediate effect of lime on the vegetation of the land to which it is applied is very striking. It immediately destroys all sorts of moss, makes a tender herbage spring up, and eradicates a number of weeds. It improves the quantity and quality of most crops, and causes them to arrive more rapidly at maturity. The extent to which it produces these effects is dependent on the form in which it is applied. When the lime is used hot, that is, immediately after it has been slaked, they are produced most rapidly and effectually; but if it has been so long exposed to the air as to absorb much of the carbonic acid it lost in burning, and has got into what is commonly called the mild state, it operates more slowly; and when it is applied as chalk, marl, or pounded limestone, its action is still more tardy. Various circumstances, which must depend upon very different considerations, must necessarily influence the farmer in the selection of one or other of these different forms of lime; but on the whole, it will be found that the greatest advantages are on the side of the well-burned and freshly slaked lime. The consideration of all the minuti?to be attended to, however, would carry us far beyond the limits of this work, and trench to some extent on the subject of practical agriculture.

Various kinds of refuse matters containing lime have been used in agriculture, but they are generally inferior to good lime, and not generally more economical. The most important of these is gas lime, or lime which has been used for purifying coal gas. In going through this process it absorbs carbonic acid from the gas, and consequently passes back, more or less, completely into the form of carbonate of lime. But it also takes up sulphur, which remains in it in the form of sulphuret of calcium. It is well known that all sulphurets are prejudicial to vegetable life, and hence, when fresh gas lime is used, its effects are often injurious rather than beneficial. But if it be exposed for some time to the air, oxygen is absorbed, the sulphur is converted into sulphuric acid, gypsum is produced to the extent of some per cent, and the lime then becomes innocuous. When composted with dry soil, the admission of air into the interior of the lime is facilitated, and this change takes place with greater rapidity. The waste lime from bleach-works, tanneries, and other manufactories, is occasionally used by farmers; but unless obtained at a nominal price, it cannot compete with good quick lime, owing to the large amount of water it contains, and the consequent increase in the cost of carriage.

Sulphate of Lime or Gypsum.--Gypsum has been extensively used as a manure, and is found to exert a very remarkable influence upon clover, and leguminous crops generally. It is employed in quantities varying from two cwt. per acre up to a very large quantity, and almost invariably with good results, in some instances even with the production of double crops. Much speculation has taken place as to the cause of this action which is so specific in its character, and from Sir Humphrey Davy down to the present time, many chemists and agriculturists have considered the matter. Sir Humphrey Davy attributed its action to its supplying sulphur to those plants which, according to him, contain an unusually large quantity of that element. That opinion has been since entertained by others, but it can scarcely be considered as well founded, for the more accurate experiments recently made do not point to any conspicuous differences between the quantities of sulphur contained in these and other plants. It is, moreover, to gypsum alone that these effects are due, and if it were merely as a source of sulphur that it was employed, there are other salts which could be equally, perhaps more advantageously, used; such, for instance, as sulphate of soda. Others have attributed its action to its power of fixing ammonia, but this explanation is certainly untenable, for the soil itself possesses this property very powerfully, and it is inconceivable that the addition of a few hundred weights of gypsum should have any effect in promoting this action. The experiments which have been made with gypsum leave no doubt as to its effect, more especially on leguminous plants, but they do not afford an explanation of its mode of action, for which further inquiries, directed especially to that object, are required.

The application of gypsum to the soil appears to have diminished of late years, and this is probably due to the large consumption of superphosphates, and other manufactured manures, which contain it in abundance. In an ordinary application of these substances, there are contained from one to two hundredweight of gypsum; and it is not likely that when they have been extensively used, much benefit will be derived from a further application of it by itself.

CHAPTER XII.

THE VALUATION OF MANURES.

The determination of the value of a manure is in many respects a commercial rather than a chemical question, but as it must be founded on the analysis, and presents some peculiarities dependent on the complicated nature of the substances to be valued, it has fallen to some extent into the hands of the chemist. The principle on which the value of any commercial sample is estimated is very simple. It is only necessary to know the price of the pure article, and that of the particular sample to be valued is obtained by making a deduction from this price proportionate to the per centage of impurities shewn by the analysis. Thus, for example, if pure sulphate of ammonia sells at ?6 per ton, a sample containing 10 per cent of impurities ought to be purchased for ?4: 8s., and so on for any other quantity. This system which answers perfectly with sulphate of ammonia, nitrate of soda, or any other substance whose value depends on one individual element, is inapplicable in the case of complex manures, such as guano and the like, in which several factors combine to make up the value. In such cases, manures of very different composition may have the same value, the deficiency in one particular element being counterbalanced by the excess of another. Hence it becomes necessary to obtain an estimate of the value of each factor, from which that not only of one particular substance, but of every possible mixture may be determined.

When we come to inquire minutely into this question, it appears that the commercial value of any substance is not estimated solely by considerations of composition, but is dependent to a great extent on questions of demand and supply, and applicability to particular purposes. Thus coprolites containing from 55 to 60 per cent of phosphates sell at about ?: 12s. per ton, while bone-ash containing the same quantity of that ingredient brings about twice as much; in other words, phosphates are nearly twice as valuable in bone-ash as in coprolites, and as a phosphatic guano their price is generally still higher; and the reason for this is obvious, in bones and guano the phosphates are in a high state of division, in which they are easily attacked and disintegrated by the carbonic acid of the soil, and rendered available to plants; while in coprolites they are in a hard and compact form, and are of little use unless they have previously undergone an expensive preparation. In the same way, if the market price of different kinds of guano be inquired into, very great differences are found to exist in the rate at which phosphates are sold, and this is attributable in part to the fact that the price at which any article is charged commercially, is such as to cover the prime cost, expense of freight, and other charges, and to leave a profit to the importer; and partly, also, no doubt, to the carelessness with which manures are often purchased, and to the want of careful field experiments in which the effects produced by them are properly compared. It will be readily understood that the state of division of any substance, the readiness with which its constituents can be rendered available to the plants, care of application, and many other circumstances must influence its price; but making due allowance for these, differences are met with which appear to some extent to be merely the result of caprice. It is easy to understand why bone-ash should sell at double the price of coprolites, but no good reason can be shewn why the phosphates in one kind of guano should be sold at a much higher price than another, and the difference would probably disappear if greater attention were paid to the results of field experiments.

However great and inexplicable these differences may be, it is not the business of the valuator of a manure to discuss them. On the contrary, he is bound to accept them as the basis of his calculation, and to endeavour to deduce from them a proper system of estimation for each substance. Strictly speaking, each individual manure ought to be valued according to a plan special to itself, and deduced from its own standard market price; but it is obvious that this would lead to innumerable complications and defeat its own ends, and hence an attempt has been made to contrive a general system suited to all manures, and which, though not absolutely correct, is a sufficient approximation for all practical purposes, and a tolerably accurate guide to the determination of their relative values.

The constituents of a manure which are of actual value are ammonia, insoluble phosphates, biphosphate of lime (soluble phosphates), sulphate of lime, nitric acid (as nitrate of soda), potash, soda, and organic matter. These substances differ greatly in value. Ammonia and phosphates, soluble and insoluble, are costly; and by far the larger part of the value of all guanos, and the common manufactured manures, depends on them. Nitric acid and potash are also very valuable substances, but as they are rarely found in manufactured manures, and never in sufficient quantity to exert any material influence in their price, it is not usual to take them into consideration except in particular cases. The alkali which commonly exists in artificial manures is soda, and when alkaline salts appear in any analysis, they must be assumed to consist almost entirely of that substance generally in the form of common salt, and be valued accordingly. Sulphate of lime and organic matter though abundant constituents of most manures, add but little to their value, and it is a moot point whether they ought to be taken into consideration, although most persons allow a small value for them. Carbonate of lime, sand, or siliceous matter, and water, of course, are altogether worthless.

In order to obtain the value of a manure containing several of these substances, it is necessary to ascertain the average commercial price of each individually. This is easily done when they are met with in commerce separately, or at least mixed only with worthless substances, but some of them are only found in complex mixtures, and in these cases it is necessary to arrive at a result by an indirect process, according to methods which will be immediately explained. The question to be solved is the price actually paid for a ton of each substance in a pure state, and we shall proceed to consider them in succession.

Insoluble Phosphates.--These are purchased alone, chiefly in the form of coprolites and bone-ash, or the spent animal charcoal of the sugar refiners. Ground coprolites, containing about 58 per cent of phosphates, sell at ?: 12s. per ton, which is at the rate of ?: 8s. for pure phosphates. Bone-ash varies considerably in price, but of late samples containing 70 per cent of phosphates have sold as low as ?: 10s. per ton, and consequently pure phosphates in this form are worth ?: 8s. per ton. Although these are the only forms in which phosphates are purchased alone, it is possible to determine the price at which they are sold in bones and phosphatic guanos, by first deducting the value of the ammonia they contain, and assuming the remainder to represent the price paid for the phosphates. In this way we find the following values for insoluble phosphates:--

In Coprolites ? 10 0 Bone-ash 6 8 0 Bones 7 5 0 Phosphatic guanos 10 0 0

It is to be observed that these are actual prices, and they are liable to fluctuate with the state of the market, although they are pretty fair averages. It is important to notice how much they vary in the different forms; the farmer who buys a phosphatic guano paying for phosphates a much higher price than he could have obtained those for in other substances--a difference which must be attributed to the high state of division in which they exist in the guano. We do not here enter upon the question how far this difference in price is justified; we are content with the fact that it exists, and we are compelled to estimate the value of phosphates in a phosphatic guano at the price given above, although in Peruvian guano they are sold at a lower rate. For all other manures, of which bones and bone-ash form the basis, ? may be taken as a fair price, and it is that usually adopted, though ? and ?0 have sometimes been assumed as the average.

Ammonia is met with in commerce as muriate and sulphate of ammonia.

The former, owing to its high price, is practically excluded from use as a manure; the latter sells at present at from ?5 to ?5: 10s. per ton, and, making allowance for the usual amount of impurity (5 or 6 per cent), the actual ammonia is worth about ?3 per ton. Calculating from other substances it appears that ammonia is worth, per ton, in--

Sulphate of ammonia ?3 0 0 Bones 61 0 0 Peruvian guano 57 0 0

the average being ?0, which is the price usually adopted.

Sulphate of Lime and Alkaline Salts (consisting chiefly of soda) are generally estimated at ? per ton; and potash in those cases, in which it is necessary to take it into account, is usually valued at from ?0 to ?0 per ton, the former being its value in kelp, the form in which it can be most cheaply purchased.

Nitrate of Soda is usually sold at from ?5 to ?5: 10s. per ton, and, making allowance for impurities, ?6 may be taken as the value of the pure salt.

Biphosphate of Lime, Soluble Phosphates.--Considerable difficulty is experienced in estimating the value of these substances, because they are not met with in commerce alone, or in any form except that of superphosphate, and the prices at which they are sold in different samples of that manure differ excessively. The only course by which any result can be obtained, is to determine the average price of a good superphosphate, and putting the values already ascertained on all the other constituents to reckon the difference between that sum and the market price as the value of soluble phosphates. Throwing out, as inferior, all samples containing less than 10 per cent of soluble phosphates, and taking the good only, I find that the average composition of the phosphates in the market during the present year has been--

Water 10?1 Organic matter 9?3 Biphosphate of lime equivalent to 19?3 "soluble phosphates" 12?5 Insoluble phosphates 14?8 Sulphate of lime 45?4 Alkaline salts 2?1 Sand 5?8 ------ 100?0 Ammonia 1?1

It is more difficult to fix the average price of superphosphate, as in many cases no information could be obtained on this point; but among those analyzed were samples at all prices, from ? up to ?0: 10s. per ton, so that on the whole, ? may be assumed as an average, and in that case soluble phosphates are worth ?7: 19s. per ton. Had the inferior samples been included, the price would have been higher, and in fact the rate at which soluble phosphates have been commonly estimated is ?0 per ton, or ?6: 16s. for biphosphate of lime, although sometimes the former have been reckoned as low as ?5, with a corresponding rate for the latter. It is important that biphosphate of lime and soluble phosphates should not be confounded with one another in valuing a manure, the latter having one and a half times the value of the former.

As manures are liable to considerable fluctuations in price, the value attached to each of their constituents ought to be varied with the state of the market; but it is obviously impossible for the farmer to watch the changes in price with such minuteness as to enable him to do this, and it is much more convenient, as well as safer, to adopt a fixed average, which can be used with reasonable accuracy at all times. The fact is, that this system of valuation is only an approximation to the truth; and if absolute accuracy were aimed at, it would be necessary to vary the estimates, not only at different times, but at different localities at the same time, and to some extent also according to the kind of manure. The price of soluble phosphates more especially, fluctuates to a great extent, being practically fixed by each manufacturer according to the facilities which his position or command of raw material offer for producing them at a low rate. We thus find that when made from bones alone, the cost of that substance is not unfrequently as high as ?0 per ton, and when bone-ash alone is used it is sometimes as low as ?0. Such extreme differences, of course, cannot be taken into account in the system of valuation adopted, where all that can be done is to take average values, which, when applied to average samples, ought to bring out their value.

The data which have already been given regarding the price of the individual constituents of manures can be applied to the determination of the value of any mixture in two different ways by means of the subjoined table:--

	Price per Ton.	Per cent per Ton.
Ammonia	?0 0 0	? 12 0
Insoluble phosphates	7 0 0	0 1 5
Do. in phosphatic guanos	10 0 0	0 2 0
Soluble phosphates	30 0 0	0 6 0
Biphosphate of lime	46 16 0	0 9 4-1/2
Alkaline salts	1 0 0	0 0 2-4/10

| | Sulphate of lime | 1 0 0 | 0 0 2-4/10 | | Potash | 20 0 0 | 0 4 0 | | Nitrate of soda | 16 0 0 | 0 3 2-1/2 | | Organic matter | 0 10 0 | 0 0 1-1/4 | +-----------
------------------+------------------+--------------------+

Supposing it be desired to calculate the value of a manure by the first column, it is obvious that if we suppose 100 tons to be purchased, the per centages of the different constituents shewn in the analysis will give the number of tons of each contained in 100 tons of the mixture, and, selecting the analysis of the superphosphate given in a previous page, we proceed in the calculation as follows:--

14?1 tons of organic matter at 10s. ? 0 0 14?6 " soluble phosphates at ?0 446 0 0 15?3 " insoluble phosphates at ? 105 0 0 39?3 " sulphate of lime at ? 39 0 0 3?2 " alkaline salts at ? 4 0 0 2?0 " ammonia at ?0 126 0 0 ----------
Value of 100 tons ?27 0 0 or ? : 5s. per ton.

According to the second column, the numbers give the sum by which the per centages of each ingredient must be multiplied, to give its value in a ton of manure, and it is used for the same manure in the following manner:--

14?1 organic matter, multiplied by 1-1/4d. ? 1 5 14?8 soluble phosphates " 6s. 4 9 2 15?3 insoluble phosphates " 1s. 5d. 1 1 4 39?3 sulphate of lime " 2-4/10d. O 8 10 3?2 alkaline salts " 2-4/10d. O O 9 2?0 ammonia " 12s. 1 5 3 -------- Value per ton ? 6 9

The difference is due to the less minute calculation of fractional quantities in the latter case.

The calculation of the value of any other manure is effected in exactly the same manner, taking care, however, to use the higher value for phosphates in the case of a phosphatic guano. It will be obvious to every one who tries the two methods that the first greatly exceeds the second in convenience and simplicity in the calculations, and it is that most commonly in use, although some persons prefer the second.

Although the data just given must always form the basis of the valuation of any manure, there are a variety of other circumstances which must be taken into account, and which give great scope for the judgment and experience of the valuator. Of these the most important is the proper admixture of the ingredients, and the condition of the manure as regards dryness, complete reduction to the pulverulent state, and the like. A certain allowance ought always to be made for careful manufacture; and, on the other hand, where the manure is damp or ill reduced, a small deduction (the amount of which must be decided by the experience of the valuator) ought to be made on account of the risk which the farmer runs of loss from unequal distribution, and the extra cost of carriage of an unnecessary quantity of water.

It is also necessary to take into account the particular element required by the soil. Thus, a farmer who finds his soil wants phosphates, will look to the manure containing the largest quantity of that substance, and possibly not requiring ammonia, will not care to estimate at its full value any quantity of that substance which he may be compelled to take along with the former, but will look only to the source from which he can obtain it most cheaply. It may be well, therefore, to point out that ammonia is most cheaply purchased in Peruvian guano; insoluble phosphates in coprolites; and soluble phosphates in superphosphates, made from bone-ash alone. In general, however, it will be found most advantageous to select manures in which the constituents are properly adjusted to one another, so that neither ammonia, soluble nor insoluble phosphates, preponderate; but, of course, it must frequently happen that it will prove more economical to buy the substances separately and to make the mixture, than to take the manure in which they are ready mixed.

In judging of the value of any manure, it is also important to make sure that the analysis which forms the basis of the calculation is that of a fair sample, which correctly represents the bulk actually delivered to the purchaser, and not one which has been made to do duty for an unlimited quantity of manure, which is supposed to be all of equal quality, as often happens in the hands of careless manufacturers, and too great attention cannot be devoted to the selection of the sample, which is very often done in an exceedingly slovenly manner.

CHAPTER XIII.

THE ROTATION OF CROPS.

Reference has already been more than once made to the fact that a crop growing in any soil must necessarily exhaust it to a greater or less extent by withdrawing from it a certain quantity of the elements to which its fertility is due. That this is the case has been long admitted in practice, and it has also been established that the exhausting effects of different species of plants are very different; that while some rapidly impoverish the soil, others may be cultivated for a number of years without material injury, and some even apparently improve it. Thus, it is a notorious fact that white crops exhaust, while grass improves the soil; but the improvement in the latter case is really dependent on the fact, that when the land is laid down in pasture, nothing is removed from it, the cattle which feed on its produce restoring all but a minute fraction of the mineral matters contained in their food; and as the plants derive a part, and in some instances a very large part, of their organic constituents from the air, the fertility of the soil must manifestly be increased, or at all events maintained in its previous state. When, however, the plant, or any portion of it, is removed from the soil, there must be a reduction of fertility dependent on the quantity of valuable matters withdrawn by it; and thus it happens that when a plant has grown on any soil, and has removed from it a large quantity of nutritive matters, it becomes incapable of producing an equally large crop of the same species; and if the attempt be made to grow it in successive years, the land becomes incapable of producing it at all, and is then said to be thoroughly exhausted. But if the exhausted land be allowed to lie for some time without a crop, it regains its fertility more or less rapidly according to circumstances, and again produces the same plant in remunerative quantity. The observation of this fact led to the introduction of naked fallows, which, up to a comparatively recent period, were an essential feature in agriculture. But after a time it was observed that the land which had been exhausted by successive crops of one species was not absolutely barren, but was still capable of producing a luxuriant growth of other plants. Thus peas, beans, clover, or potatoes, could be cultivated with success on land which would no longer sustain a crop of grain, and these plants came into use in place of the naked fallow under the name of fallow crops. On this was founded the rotation of crops; for it was clear that a judicious interchange of the plants grown might enable the soil to regain its fertility for one crop at the time when it was producing another; and when exhausted for the second, it might be again ready to bear crops of the first.

The necessity for a rotation of crops has been explained in several ways. The oldest view is that of Decandolle, who founded his theory on the fact that the plants excrete certain substances from their roots. He found that when plants are grown in water, a peculiar matter is thrown off by the roots; and he believed that this extrementitious substance is eliminated because it is injurious to the plant, and that, remaining in the soil, it acts as a poison to those of the same species, and so prevents the growth of another crop. But this excretion, though poisonous to the plants from which it is excreted, he believed to be nutritive to those of another species which is thus enabled to grow luxuriantly where the others failed. Nothing can be more simple than this explanation, and it was readily embraced at the time it was propounded and considered fully satisfactory. But when more minutely examined, it becomes apparent that the facts on which it is founded are of a very uncertain character. Decandolle's observations regarding the radical excretions of plants have not been confirmed by subsequent observers. On the contrary, it has been shewn that though some plants, when growing in water, do excrete a particular substance in small quantity, nothing of the sort appears when they are grown in a siliceous sand. And hence the inference is, that the peculiar excretion of plants growing in water is to be viewed as the result of the abnormal method of their growth rather than as a natural product of vegetation. But even admitting the existence of these matters, it would be impossible to accept the explanation founded upon them, because it is a familiar fact that, on some soils, the repeated growth of particular crops is perfectly possible, as, for instance, on the virgin soils of America, from which many successive crops of wheat have been taken; and in these cases the alleged excretion must have taken place without producing any deleterious effect on the crop. Besides, it is in the last degree improbable that these excretions, consisting of soluble organic matters, should remain in the soil without undergoing decomposition, as all similar substances do; and even if they did, we cannot, with our present knowledge of the food of plants, admit the possibility of the direct absorption of any organic substance whatever. Indeed, the idea of radical excretions, as an explanation of the rotation of crops, must be considered as being entirely abandoned.

The necessity for a rotation of crops is now generally attributed to the different quantities of valuable matters which different plants remove from the soil, and more especially to their mineral constituents. It has been already observed that great differences exist in the composition of the ash of different plants in the section on that subject; and it was stated that a distinction has been made between lime, potash, and silica plants, according as one or other of these elements preponderate in their ashes. The remarkable difference in the proportion of these elements has been supposed to afford an explanation of rotation. It is supposed that if a plant requiring a large quantity of any one element, potash, for example, be grown during a succession of years on the same soil, it will sooner or later exhaust all, or nearly all, the potash that soil contains in an available form, and it will consequently cease to produce a luxuriant crop. But if this plant be replaced by another which requires only a small quantity of potash and a large quantity of lime, it will flourish, because it finds what is necessary to its growth. In the meantime, the changes which are proceeding in the soil, are liberating new quantities of the inorganic matters from those forms of combination in which they are not immediately available, and when after a time the plant which requires potash is again sown on the soil, it finds a sufficient quantity to serve its purpose. We have already, in treating of the ashes of plants, pointed out the extent of the differences which exist; but these will be made more obvious by the annexed table, giving the quantity of the different mineral matters contained in the produce of an imperial acre of the different crops.

TABLE shewing the quantities of Mineral Matters and Nitrogen in average Crops of the principal varieties of Farm Produce.

Produce per Imperial Acre	Weight in lbs	Total Mineral Matters	Potash	Soda	Lime
Wheat--Grain	28 bushels at 60 lbs	1,680	34?2	10?1	1?0	1.04
Straw	1 ton 3 cwt	2,576	114?8	20?0	2?4	8?3
Total	...	...	148?0	30?1	4?4	9?7
Barley--Grain	33 bushels at 53 lbs	1,749	44?4	9?0	0?0	0?6
Straw	18 cwt	2,106	99?4	11?4	1?4	5?1
Total	...	...	143?8	20?4	1?4	6?7
Oats--Grain	34 bushels at 40 lbs	1,360	48.89	11?0	...	5?1
Straw	1 ton	2,240	143?3	30?1	6?0	10?9
Total	...	...	192?2	41?1	6?0	15?0
Beans, Peas--Grain	25 bushels at 60 lbs	1,650	55?7	30?0	0?1	3?1
Straw	1 ton	2,240	108?1	48?1	13?4	29?7
Total	...	...	164?8	78?1	13?5	32?8
Turnips--Bulbs	13-1/2 tons	30,240	213?5	57?5	44?1	28?0

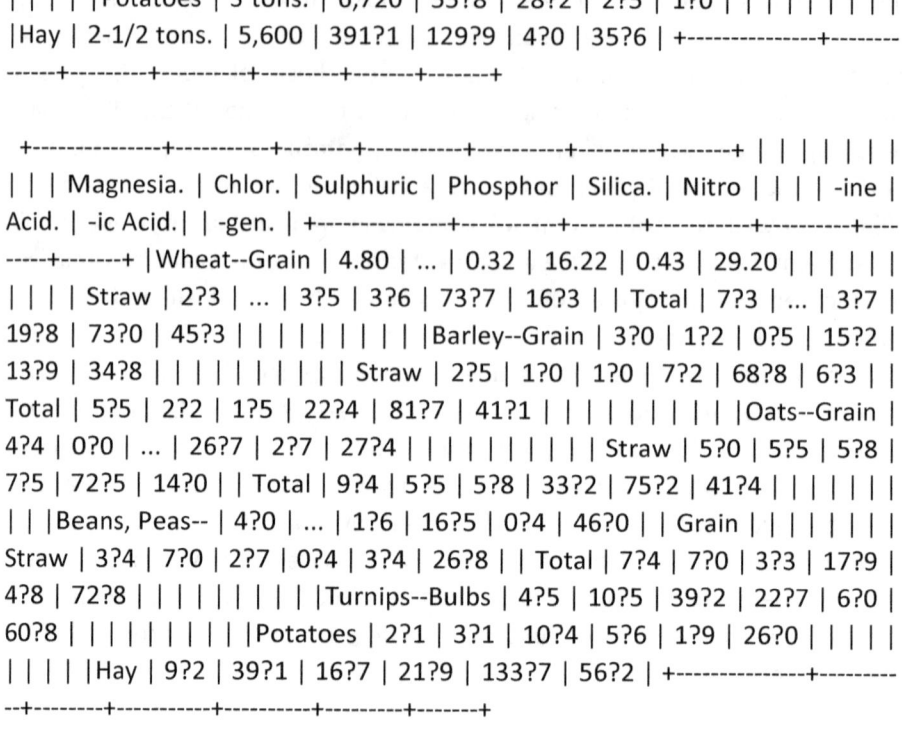

The minor constituents, such as oxide of iron, manganese, etc., have been omitted as being of little importance; and the quantity of nitrogen, which is of great moment in estimating the exhaustive effects of various crops, has been added.

In examining this table, it becomes apparent that while in regard to some of the elements, the quantities removed by different crops do not differ to any marked extent, in others the variation is very great. The cereals and grasses are especially distinguished by the larger quantity of silica they contain, and the exhaustive effect consequent upon the removal of both grain and straw from soils which contain but a limited supply of that substance in an available condition is obvious. It is clear that under such circumstances the frequent repetition of a cereal crop may so far diminish the amount of available silica as to render its cultivation impossible, although the other substances may be present in sufficient quantity to produce a plentiful crop of any plant which does not require that element. Beans and peas, turnips and hay, on the other hand, require a very large quantity of alkalies, and especially of potash.

Looking more minutely, however, into this matter, certain points attract attention which appear to be at variance with commonly received opinions. With the exception of silica, for example, the cereals do not withdraw from the soil so large a quantity of mineral matters as some of the so-called fallow crops, and if their straw be returned to the soil they are by far the least exhaustive of all cultivated plants; and we thus recognise the justice of that practical rule, which lays it down as an essential point of good husbandry that the straw ought, as far as possible, to be consumed on the farm on which it is produced. As regards the general constituents of the ash, it is also to be remarked that though differences in their proportions exist, they are by no means so marked as might be expected; thus there are no plants for which a large quantity of potash, nitrogen, and phosphoric acid is not required; and it is not very easy to see how the substitution of the one for the other should be of much importance in this respect. Indeed, the more minutely the subject is examined, the more do we become convinced of the insufficiency of that view which attributes the necessity for a rotation of crops to differences in chemical composition alone. There can be no doubt that the nature of the plant and the particular mode in which it gathers its nutriment, have a most important influence. Certain plants are almost entirely dependent on the soil for their organic constituents, while others derive a large proportion of them from the air, and a plant of the latter class will flourish in a soil in which one of the former is incapable of growing. In other cases, the structure and distribution of the roots is the cause of the difference. Some plants have roots distributed near the surface and exhaust the superficial layer of the soil, others penetrate into the deeper layers, and not only derive an abundant supply of food from them, but actually promote the fertility of the surface soil by the refuse portions of them which are left upon it. Experience has in this respect arrived at results which tally with theory, and it is for this reason that the broad-leafed turnip, which obtains a considerable quantity of its nutriment from the air, alternates with grain crops which are chiefly dependent on the soil. It is undoubtedly to some such cause that several remarkable instances of what may be called natural rotations are to be attributed. It is well known in Sweden that when a pine forest is felled, a growth, not of pine but of birch, immediately springs up. Now the difference in composition of the ash of these trees is not sufficient to explain this fact, and it must clearly be due to some difference in the distribution of their roots, or the mode in which they obtain their food.

Whatever weight may be given to these different explanations of rotation, there is no doubt about the importance of attending to it, and there are various practical deductions of much importance to be drawn from the facts with which we are acquainted. Thus it is to be observed that the quantities of mineral matters withdrawn by plants of the same class are generally similar, and hence it may be inferred that crops of the most opposite class ought as much as possible to alternate with one another, and each plant should be repeated as seldom as possible, so that, even when it is necessary to return to the same class, a different member of it should be employed. Thus, for instance, in place of immediately repeating wheat, when another grain crop is necessary, it would theoretically be preferable to employ oats or barley, and to replace the turnip by mangold-wurzel or some other root. It is obvious, however, that this system cannot be carried out in practice to its full extent; for the superior value of individual crops causes the more frequent repetition of those which make the largest return. But experience has so far concurred with theory that it has taught the farmer the advantage of long rotations; and we have seen the successive introduction of the three, four, five, and six-course shift, and even, in some instances, of longer periods.

Such is the theory of rotation, and while it will always be most advantageous to adhere to it, it is by no means necessary that this should be done in an absolutely rigid manner. In the practice of agriculture, plants are placed in artificial circumstances, and instead of allowing them to depend entirely on the soil, they are supplied with a quantity of manure containing all the elements they require, and if it be used in sufficiently large quantity, the same crop may be grown year after year. And accordingly the order of rotation, which is theoretically the best, may be, and every day is, violated in practice, although this must necessarily be done at the expense of a certain quantity of the valuable matters of the manure added, and is so far a practice which ought theoretically to be avoided. In actual practice, however, the matter is to be decided on other grounds. The object then is, not to produce the largest crops, but those which make the largest money return, and thus it may be practically economical to grow a crop of high commercial value more frequently than is theoretically advantageous. In such cases the farmer must seek to do away as far as possible with the disadvantages which such a course entails, and this he will endeavour to accomplish by careful management and a liberal treatment of the soil.

But while this system may be adopted to some extent, it must also be borne in mind that the frequent repetition of some crops cannot be practised with impunity, for plants are liable to certain diseases which manifest themselves to the greatest extent when they have been too often cultivated in the same soil. Clover sickness, which affects the plant when frequently repeated on light soils, and the potatoe disease and finger and toe have been attributed to the same cause. Whether this is the sole origin of these diseases is questionable, but there is no doubt that they are aggravated by frequent repetition, and hence a strong argument in favour of rotation. It has been asserted by great authorities in high farming, that with our present command of manures, rotations may be done away with; but this is an opinion to which science gives no countenance, and he would be a rash man who attempted to carry it out in practice.

CHAPTER XIV.

THE FEEDING OF FARM STOCK.

The feeding of cattle, once a subordinate part of the operations of the farm, has now become one of its most important departments, and a large number of minute and elaborate experiments have been made by chemists and physiologists with the view of determining the principles on which its successful and economical practice depends. These investigations, while they have thrown much light on the matter, have by no means exhausted it, and it will be readily understood that the complete elucidation of a subject of such complexity, touching on so many of the most abstruse and difficult problems of chemistry and physiology, and in which the experiments are liable to be affected by disturbing causes, dependent on peculiarities of constitution of different animals, cannot be otherwise than a slow process.

In considering the principles of feeding, it is necessary to point out, in the first instance, that the plant and animal are composed of the same chemical elements, hence the food supplied to the latter invariably contains all the substances it requires for the maintenance of its functions. And not only is this the case, but these elements are to a great extent combined together in a similar manner,--the fibrine, caseine, albumen, and fatty matters contained in animals corresponding in all respects with the compounds extracted from plants under the same name; and though the starchy and saccharine substances do not form any part of the animal body, they are represented in the milk, the food which nature has provided for the young animal. It has been frequently assumed that the nitrogenous and fatty matters are simply absorbed into the animal system, and deposited unchanged in its tissues; but it is probable that the course of events is not quite so simple, although, doubtless, the decomposition which occurs is comparatively trifling. The starchy matters, on the other hand, are completely changed, and devoted to purposes which will be immediately explained.

It is a matter of familiar experience, that if the food be properly proportioned to the requirements of the animal, its weight remains unchanged; and the inference to be drawn from this fact obviously is, that the food does not remain permanently in the system, but must be again got rid of. It escapes partly through the lungs, and partly by the excretions, which do not consist merely of the part which has not been digested, but also of that portion which has been absorbed, and after performing its allotted functions within the system, has become effete and useless. When the weights of the excretions, the carbon contained in the carbonic acid expired by the lungs and the small quantity of matter which escapes in the form of perspiration, are added together, they are found in such a case to be exactly equal to the food. If the animal be deprived of nutriment, it immediately begins to lose weight, because its functions must continue--carbon must still be converted into carbonic acid to maintain respiration--and the excretions be eliminated, although diminished in quantity, because they no longer contain the undigested portion of the daily food, and the substances already stored up in the body are consumed to maintain the functions of life. Universal experience has shewn that, under such circumstances, the fat which has accumulated in various parts of the body disappears, and the animal becomes lean; but it is less generally recognised that the muscular flesh, that is the lean part of the body, also diminishes, although it is sufficiently indicated by the fact that nitrogen still continues to be found in the urine, and that the animal becomes feeble and incapable of muscular exertion. Respiration and secretion, in fact, proceed quite irrespective of the food, which is only required to repair the loss they occasion. When the course of events within the animal body is traced, it is found to be somewhat as follows: The food consumed is digested and absorbed into the blood, where it undergoes a series of complicated changes, as a consequence of which part of it is converted into carbonic acid, and eliminated by the lungs, and part is deposited in the tissues as fat and flesh. After the lapse of a certain period, longer or shorter according to circumstances, a new set of actions comes into play, by which the complex constituents of the tissues are resolved into simpler substances, and excreted chiefly by the lungs and kidneys. The changes thus produced are, to a great extent, identical with those which would take place if the fat and flesh were consumed in a fire; and the animal frame may, in a certain sense, be compared to a furnace, in which, by the daily consumption of a certain quantity of fuel and air inhaled in the process of respiration, its temperature is maintained above that of the surrounding atmosphere. If the daily supply of fuel, that is of food, be properly adjusted to the loss by combustion, the weight of the animal remains constant; if it be reduced below this quantity, it diminishes; but if it be increased, the stomach either refuses to digest and assimilate the excess, or it is absorbed and stored up in the body, increasing both the fat and flesh.

When an animal is fed in such a manner that its weight remains constant, a balance is produced between the supply of nutriment contained in the food and the waste of the tissues, the gain from the former exactly counterpoising the loss occasioned by the latter. If in this state of matters an additional supply of food be given, this balance is deranged, and the nutriment being in excess of the loss, the animal gains weight, and it continues to do this for some time, until it reaches a point at which a new balance is established, and its weight again becomes constant; and this is due to the fact that the animal becomes subject to an additional waste, consequent on the increased weight of matter accumulated in its tissues. If, after the animal has attained its new constant weight, the food be a second time increased, a further gain is obtained, and so on, with every addition to the supply of nutriment, until at length a certain point is reached, beyond which its weight cannot be forced. In fact, each successive increase of weight is obtained at a greater expenditure of food. If, for example, a lean animal is taken, and its food increased by a given quantity, it will rapidly attain a certain additional weight, but if another extra supply of food be given, the increase due to it will be much more slowly attained, and so on until at length an additional increase can only be secured by the long-continued consumption of a very large quantity of food. The great object of the feeder is to obtain the greatest possible increase with the smallest expenditure of food, and to know the point beyond which it is no longer economical to attempt to force the process of fattening. To do this it is necessary first to consider the composition of the animal itself, then that of its food, and lastly, the mode in which it may be most economically used.

It has been already observed that the animal tissues are composed of albuminous or nitrogenous compounds, fat, mineral matters, and water; but the proportions of these substances have, until lately, been very imperfectly known. Water is well known to be by far the largest constituent, and amounts in general to about two-thirds of the entire weight, and it has been generally supposed that the nitrogenous matters stood next in point of abundance, but a most important and elaborate series of experiments by Messrs. Lawes and Gilbert have shewn that they are greatly exceeded by the fatty matters. The following table contains a summary of the composition of ten different animals in different stages of fattening. The first division gives the composition of the carcass, that is, the portion of the animal usually consumed as human food; the second that of the offal, consisting of the parts not usually employed as food; and the third that of the entire animals, including the contents of the stomach and intestines:--

[Transcriber's note: Column titles are printed vertical, which is not possible to do here. Therefore they are replaced with a 2-3 character code, explained here]

Column titles: MM = Mineral Matter NC = Nitrogenous Compounds TDS = Total Dry Substance CSI = Contents of Stomachs and Intestine in moist state. Wat = Water

	Per cent in Carcass					Per cent in Offal, excluding Intestines.					Per cent in entire animals, including contents of Stomachs and Intestines			
	MM	NC	Fat	TDS	WAT	MM	NC	Fat	TDS	WAT	MM	NC	Fat	TDS	WAT
Fat Calf	4?8	16?	16?	37?	62?	3?1	17?	14?	35?	64?					
Half-fat Ox	5?6	17?	22?	46?	54?	4?5	20?	15?	40?	59?					
Fat Ox	4?6	15?	34?	54?	45?	3?0	17?	26?	47?	52?					
Fat Lamb	3?3	10?	36?	51?	48?	2?5	18?	20?	41?	58?					
Store Sheep	4?6	14?	23?	42?	57?	2?9	18?	16?	36?	63?					
Half-fat old Sheep	4?3	14?	31?	50?	49?	2?2	17?	18?	38?	61?					
Fat Sheep	3?5	11?	45?	60?	39?	2?2	16?	26?	44?	55?					
Extra fat Sheep	2?7														

9?|55?|67?|33?|| 3?4| 16?| 34?| 54?| 45?| |Store Pig |2?7|14?|28?|44?|55?|| 3?7| 14?| 15?| 32?| 67?| |Fat Pig |1?0|10?|49?|61?|38?|| 2?7| 14?| 22?| 40?| 59?| |------------------|----|---- |----|----|----||-----|-----|-----|-----|-----| |Mean of all |3?9|13?|34?|51?|48?|| 3?2| 17?| 21?| 41?| 58?| |------------------|----|---- |----|----|----||-----|-----|-----|-----|-----| |Mean of 8, viz, the| | | | | | || | | | | | | half-fat, fat, and|3?5|13?|36?|53?|46?|| 3?2| 17?| 22?| 42?| 57?| | very fat animals | | | | | || | | | | | |------------------|----|----|----|----|----||----- -|-----|-----|-----|-----| |Mean of 6, viz., | | | | | || | | | | | | of the fat and |3?8|12?|39?|55?|44?|| 3?3| 16?| 24?| 44?| 56?| | very fat animals| | | | | || | | | | | |------------------|----|----|----|----|----||-----|-----|-----|-----|-----|

|---| | | | | | Per cent in Entire Animal.| | | | | |------------------------------| | | MM | NC | Fat| TDS| CSI | WAT| |------------------|----|----|----|----|-----|----| |Fat Calf |3?0|15?|14?|33?| 3?7|63?| |Half-fat Ox |4?6|16?|19?|40?| 8?9|51?| |Fat Ox |3?2|14?|30?|48?| 5?8|45?| |Fat Lamb |2?4|12?|28?|43?| 8?4|47?| |Store Sheep |3?6|14?|18?|36?| 6.00|57?| |Half-fat old Sheep |3?7|14?|23?|40?| 9?5|50?| |Fat Sheep |2?1|12?|35?|50?| 6?2|43?| |Extra fat Sheep |2?0|10?|45?|59?| 5?8|35?| |Store Pig |2?7|13?|23?|39?| 5?2|55?| |Fat Pig |1?5|10?|42?|54?| 3?7|41?| |------------------|----|----|----|----|-----|----| |Mean of all |3?7|13?|28?|44?| 6?3|49?| |------------------|----|----|----|----|-----|----| |Mean of 8, viz, the| | | | | | | half-fat, fat, and|3?3|13?|29?|46?| 6?6|47?| | very fat animals | | | | | | | |------------------|----|----|----|----|-----|----| |Mean of 6, viz., | | | | | | | of the fat and |3?0|12?|32?|48?| 5?8|46?| | very fat animals| | | | | | | |------------------|----|----|----|----|-----|----|

From this table it appears that, in the carcass, the proportion of fat is, in general, even in lean animals, much greater than that of nitrogenous compounds. In one case only, that of the fat calf, are they equal. But in the lean sheep there is more than one and a half times as much fat as nitrogenous matters, in the half fat sheep twice, in the fat sheep four times, and in the very fat sheep about six times as much. As a general result of the analyses it may be stated, that in the carcass of an ox in good condition, the quantity of fat will be from two to nearly three times as great as that of the so called albuminous compounds; in a sheep three or four times, and in the pig four or five times as great. In the offal, including the hide, intestines, and other parts not usually consumed as food, the proportion is very different,-- the quantity of fat being much smaller, and that of nitrogenous compounds considerably larger.

Taking a general average of the whole, the following may be assumed as representing approximately the general composition of a lean and a fat animal:--

	Lean.	Fat.
Mineral matters	5	3
Nitrogenous compounds	15	12?
Fat	24	33
Water	56	48?
	100	100?

The data given in the preceding table, coupled with a knowledge of the relative weights of the lean and fat animals, enable us to ascertain the composition of the increase during the fattening process. It is obvious, from the material diminution of the per centage of water, that the matters deposited in the tissues must contain a much larger proportion of dry matters than the whole body; and the reduced per centage of nitrogenous matters shews that the fat must also greatly preponderate. This is still more distinctly illustrated by the following table, giving the per centage composition of the increase in fattening oxen, sheep, and pigs:--

	Mineral Matters.	Nitrogenous Compounds.	Fat.	Water.
Oxen	1?7	7?9	66?	24?
Sheep	2?4	7?3	70?	20?
Pigs	0?6	6?4	71?	22?

Hence it may be stated in round numbers, that for every pound of nitrogenous matters added to the weight of a fattening animal, it will gain ten pounds of fat, and three of water. These are the proportions over the whole period of fattening, but it is probable that during the last few weeks of the process the ratio of fat to nitrogenous matters is still higher.

In considering the composition of the food of animals, it will be readily admitted that the milk, the nutriment supplied by nature for the maintenance of the young animal, must afford special instruction as to its requirements during the early stages of existence, and indicate, at least, some of the points to be attended to under the altered conditions of mature life. The following table gives the average composition of the milk of the most important farm animals:--

Cow. Ewe. Goat. Caseine 3? 4?0 4?2 Butter 3? 4?0 3?2 Sugar of milk 6? 5?0 5?8 Ash 0? 0?8 0?8 Water 86? 85?2 86?0 ------ ------ ------ 100?0 100?0 100?0

In examining these, and all other analyses of food, it is necessary to draw a distinction between the flesh-forming and the respiratory elements; the former including the nitrogenous compounds which are used in the production of flesh, the latter, the non-nitrogenous substances which produce fat and support the process of respiration. The former, however much they may differ in name, are nearly or altogether identical in chemical composition, the latter embracing two great classes--the fats which exist in the body and the saccharine compounds, including the different kinds of sugar and starch which are not found in the animal tissues. It was at one time supposed that these substances were entirely consumed in the respiratory process, and eliminated by the lungs in the form of carbonic acid and water, but it has been clearly shewn that they may be and often are converted into fat, and accumulated in the system. Careful experiments on bees have demonstrated that when fed on sugar they continue to produce wax, which is a species of fat, and animals retain their health and become fat, even when their food contains scarcely any oil. There is, however, an important difference between these two classes of substances as regards their fat-producing effect. A pound of fat contained in the food is capable of producing the same quantity within the animal; but the case is different with starch and sugar, the most trustworthy experiments shewing that two and a half pounds of these substances are necessary for that purpose. Hence we talk of the fat equivalent of sugar, by which is meant the amount of fat it is capable of producing, and which is obtained by dividing its quantity by 2?. Applying this principle to the analyses of the milk, it appears that the relative proportions of the two great classes of nutritive substances stand thus:--

Flesh Respiratory, expressed in forming their fat equivalent

Cow 3? 6? Ewe 4? 6? Goat 4? 5?

Taking the general average, it may be stated, that for every pound of flesh-forming elements contained in the food of the sucking animal, it consumes respiratory compounds capable of producing one and a half pounds of fat, and this does not differ materially from the ratio subsisting between these substances in the lean animal. When the young animal is weaned, it obtains a food in which the ratio of nitrogenous to respiratory elements is maintained nearly unchanged; but the latter, in place of containing a large amount of fatty matters, is in many cases nearly devoid of these substances, and consists almost exclusively of starch and sugar, mixed most commonly with a considerable quantity of woody fibre.

A very large number of analyses of different kinds of cattle food have been made by chemists, but our information regarding them is still in some respects imperfect. The quantity of nitrogenous compounds and of oil has been accurately ascertained in almost all, but the amount of starch, sugar, and woody fibre is still imperfectly determined in many substances. This is due partly to the fact that the nitrogenous and fatty matters were formerly believed to be of the highest importance, and might be used as the measure of the nutritive value of food to the exclusion of its other constituents, and partly also to the imperfect nature of the processes in use for obtaining the amounts of woody fibre, starch, and sugar. These difficulties have now, to a certain extent, been overcome, and the quantity of fibre and of respiratory elements has been ascertained, and is introduced, so far as is known, in the subjoined table:--

TABLE giving the Composition of the Principal Varieties of Cattle Food.

Note.--Where a blank occurs in the oil column, the quantity of that substance is so small as to be unimportant. When the respiratory elements and fibre have not been separated, the sum of the two is given.

	Nitrogenous Compounds pounds.	Oil.	Respiratory Compounds pounds.	Fibre.	Ash.	Water.
Decorticated earth-nut cake	44?0	8?6	19?4	5?3	14?5	8?2
Decorticated cotton cake	41?5	16?5	16?5	8?2	8?5	9?8
Poppy cake	34?3	11?4	23?5	11?3	13?9	6?6
Teel or sesamum cake	31?3	12?6	21?2	9?6	13?5	10?8
Rape cake	29?5	8?3	38?2	7?0	8?5	6?5
Dotter cake	29?0	7?9	27?4	16?2	12?9	

7?6 | |Tares, home-grown | 28?7 | 1?0| 58?4 | 2?0| 8?9 | |Linseed cake | 28?3 | 12?7| 35?8 | 6?2 | 6?1| 10?9 | |R 䉷 sen cake | 26?7 | 11?0| 31?7 | 16?5 | 8?0| 5?1 | |Tares, foreign | 26?3 | 1?9| 53?4 | 2?4| 15?0 | |Earth-nut cake (entire seed)| 26?1 | 12?5| 45?9 | 3?9| 11?6 | |Niger cake | 25?4 | 6?8| 42?8 | 11?5 | 8?2| 6?3 | |Beans (65 lbs. per bushel) | 24?0 | 1?9| 54?1 | 3?6| 15?4 | |Lentils | 24?7 | 1?1| 58?2 | 2?9| 12?1 | |Linseed | 24?4 | 34?0| 30?3 | 3?3| 7?0 | |Grey peas | 24?5 | 3?0| 57?9 | 2?2| 11?4 | |Foreign beans | 23?9 | 1?1| 59?7 | 3?4| 12?1 | |Cotton cake (with husk) | 22?4 | 6?7| 36?2 | 16?9 | 6?2| 11?6 | |Pea-nut cake | 22?5 | 7?2| 30?5 | 26?7 | 3?1| 9?0 | |Sunflower cake | 21?8 | 8?4| 19?5 | 33?0 | 9?3| 8?0 | |Hempseed cake | 21?7 | 7?0| 22?8 | 25?6 | 15?9| 7?1 | |Kidney beans | 20?6 | 1?2| 62?6 | 3?6| 13?0 | |Maple peas | 19?3 | 1?2| 63?8 | 2?4| 13?3 | |Madia sativa (seed) | 18?1 | 36?5| 34?9 | 4?3| 6?2 | |Clover hay (mean of | | | | | | |different species of clover)| 15?1 | 3?8| 34?2 | 22?7 | 7?9| 16?3 | |Rye | 14?0 | ...| 81?1 | 2?7 | 1?2| 14?6 | |Bran | 13?0 | 5?6| 61?7 | 6?1| 12?5 | |Oats | 11?5 | 5?9| 57?5 | 9?0 | 2?2| 13?9 | |Fine barley dust | 11?9 | 2?2| 71?1 | 2?7| 11?1 | |Wheat | 11?8 | ...| 73?2 | 0?8 | 0?2| 13?0 | |Bere | 10?5 | ...| 62?5 | 10?8 | 2?0| 14?2 | |Hay (mean of different | | | | | | | grasses) | 9?0 | 2?6| 38?4 | 29?4 | 5?4| 14?0 | |Barley | 8?9 | ...| 64?2 | 9?7 | 2?2| 14?0 | |Coarse barley dust | 8?6 | 3?7| 69?3 | 7?1| 11?3 | |Rice dust | 8?8 | 2?5| 69?2 | 8?2| 11?3 | |Oat dust | 6?2 | 3?1| 72?6 | 7?0| 9?1 | |Winter bean straw | 5?1 | ...| 67?0 | 6?9| 20?0 | |Carob bean | 3?1 | 0?1| 62?1 | 18?0 | 2?0| 12?7 | |Potato | 2?1 | ...| 17?0 | 1?7 | 1?3| 77?9 | |Carrot | 1?7 | ...| 7?1 | 3?7 | 1?1| 86?4 | |Wheat straw | 1?9 | ...| 31?6 | 45?5 | 7?7| 14?3 | |Barley straw | 1?8 | ...| 39?8 | 39?0 | 4?4| 14?0 | |Oat straw | 1?3 | ...| 37?6 | 43?0 | 4?5| 12?6 | |Mangold-wurzel | 1?4 | ...| 8?0 | 1?2 | 0?6| 87?8 | |Cabbage | 1?1 | ...| 4?3 | 1?5| 93?1 | |Turnips | 1?7 | 0?0| 4?7 | 1?8 | 1?1| 91?7 | +----------------------------+--------+------+--------+------+-------+

It is at once obvious that in many of these descriptions of food the ratio of the flesh to the fat-forming constituents differ very widely from that existing in the milk, and this becomes still more apparent when the latter are represented in their fat equivalent, as is done for a few of them in the following table:--

Flesh Respiratory, expressed forming, in their fat equivalent,

Decorticated earth-nut cake 44? 16? Linseed cake 28? 26? Tares 26?3 18? Clover hay 15?1 16? Oats 11?5 28? Hay (mean of grasses) 9?0 17? Potato 2?1 6? Wheat straw 1?9 12? Turnip 1?7 1?

It is especially note-worthy that those varieties of food, which common experience has shewn to promote the fattening of stock to the greatest extent, contain in many instances the smallest quantity of respiratory or fat-forming elements relatively to their nitrogenous compounds. This is especially the case with the different kinds of oil cake, the leguminous seeds, clover, hay, and turnips. On the other hand, in the grains the ratio is nearly that of one to three, or similar to that found in fat cattle; while in the straw, the excess of the respiratory elements is extremely great.

These facts appear at first sight to be completely at variance with the composition of the increase of fattening animals, as ascertained by Messrs. Lawes and Gilbert already referred to, and which have shewn that for every pound of nitrogenous compounds, nearly ten pounds of fat are stored within the animal; and it might be supposed that those kinds of food which contain the largest relative amount of respiratory elements ought to fatten most rapidly, and should be selected by the farmer in preference to oil-cakes and similar substances. But there are other matters to be considered, dependent on the complex nature of the changes attending the absorption and assimilation of the food. It must be particularly borne in mind that only a small proportion of the food consumed is stored up within the body, and goes to increase the weight of the animal. Even in the case of the milk, in which economy in the supply of nutritive matters has been most clearly attended to by nature, a considerable proportion escapes assimilation, and in the adult animal a large amount of the food passes off with the excretions. The justice of this position is apparent when it is remembered that an ox will go on day after day consuming from a hundred weight to a hundred weight and a half of turnips, three or four pounds of bean-meal or oil-cake, and a considerable quantity of straw, although its daily increase in live weight may not exceed a couple of pounds. And in this direction a very fertile field of inquiry lies open to the agricultural experimenter; for it would be most important to determine whether there are not some substances from which the nutritive matters may not be more easily assimilated than from others, and what proportion of each is absorbable under ordinary circumstances. On this point no information has yet been obtained applicable to individual feeding substances, but the experiments of Messrs. Lawes and Gilbert have shewn the quantity of the total food, and of each of its constituents, stored up in the fattening animal, and a summary of their results is contained in the following Table:--

TABLE shewing the Amount of each Class of Constituents, stored in the increase, for 100 consumed in the Food.

	Total Dry Matters	Mineral Matters	Nitrogenous Compounds	Fat	Substance
Sheep	3?7	4?1	9?	8?6	
Pigs	0?8	7?4	21?	17?	

Hence it appears that the pig makes a better use of its food than the sheep, retaining twice as much of its solid constituents within the body, from which may be deduced the important practical conclusion, that the former must be fattened at a much smaller cost than the latter. Looking at the individual constituents, it appears that, in the sheep, less than one-twentieth of the nitrogenous compounds, and one-tenth of the non-nitrogenous substances contained in the food, remain in the body; and a knowledge of these facts tends to modify the conclusions which might be drawn from the composition of the increase in the fattening animal. Its influence may be best illustrated by a particular example. If, for instance, the increase in a sheep contained its nitrogenous and respiratory elements in the ratio of 1 to 10, it would be totally incorrect to supply these substances in the food in the same proportions. On the contrary, it would be necessary at the very least to double the proportion of the former, because one-tenth of the fat-forming elements are absorbed, and only one-twentieth of the nitrogenous.

On further consideration, also, it seems unquestionable that the quantity of the nutritive elements stored up must depend to a large extent on the nature of the food and the particular state in which they exist in it. It is probable, or at least possible, that some kinds of food may contain their nitrogenous constituents in an easily assimilable state, and their respiratory elements in a nearly indigestible condition, or vice versa, and under these circumstances their nutritive value would be below that indicated by analysis; but these points can only be determined by elaborate and long continued feeding experiments. It is well known, however, that the mechanical state of the food has a most important influence on its nutritive value. Thus, for example, the presence of a large quantity of woody fibre protects the nutritive substances from assimilation, and seeds with hard husks pass unchanged through the animal, although, so far as their composition alone is concerned, they may be highly nutritive; and the loss of a certain quantity of many varieties of food in this way is familiar to every one.

The proper adjustment of the relative quantities of the great groups of nutritive elements in the food is a matter the importance of which cannot be over-rated, for it is in fact the foundation of successful and economical feeding; and this will be readily understood if we consider what would be the result of giving to an animal a supply of food containing a large quantity of nitrogenous and a deficiency of fat-forming compounds. In such circumstances, the animal must either languish for want of the latter, or it is forced to supply the defect by an increased consumption of food, in doing which it must take into the system a larger quantity of nitrogenous compounds than would otherwise have been requisite, and in this way the other elements, which are present in abundance, are wasted, and the theoretical and practical value of a food so constituted may be very different, and it is only when the proportions of the different groups are properly attended to that the most economical result can be obtained. It can scarcely be said that the experiments yet made by feeders enable us to fix the most suitable proportion in which those substances can be employed, although experience has led them to the use of mixtures which are in most cases theoretically correct; thus they combine oil-cakes or turnips with straw, which is poor nitrogenous, and rich in fat-forming elements; and in general it will be found that where different kinds of food are mixed, the deficiencies of the one are counterbalanced by the other, and though this has hitherto been done empirically, it cannot be doubted that as our knowledge advances it will more and more be determined by reference to the composition of the food.

Although the presence of a sufficient quantity of nutritive compounds in the food is necessarily the fundamental matter for consideration, its bulk is scarcely less important. The function of digestion requires that the food shall properly fill the stomach, and however large the supply of nutritive matters may be, their effect is imperfectly brought out if the food is too small in bulk, and it actually may become more valuable if diluted with woody fibre, or some other inert substance. At first sight this may appear at variance with the observations already made as to the effects of woody fibre in protecting the nutritive matters from absorption; but practically there are two opposite evils to be contended against, a food having too small a bulk, or one containing so large a proportion of inert substances as to become disadvantageously voluminous. The most favourable condition lies between the two extremes, and the natural food of all herbivorous animals is diluted with a certain amount of woody fibre. When these are replaced by substances containing a large quantity of nutritive matters in a small bulk, the result is that the natural instinct of the animal causes it to continue feeding until the stomach is properly distended, and it consequently consumes a much larger quantity of food than it is capable of digesting, and a more or less considerable quantity passes unchanged through the intestines, and is lost. On the other hand, if the food be too bulky, the sense of repletion causes the animal to cease eating long before it has obtained a sufficient supply of nutritive matter. It is most necessary, therefore, to study the mixture of different kinds of food, so as to obtain a proper relation between the bulk and the nutritive matters contained in the mixture; and on examining the nature of the mixed foods most in vogue among feeders, it will be found that a very bulky food is usually conjoined with another of opposite qualities. Hence it is that turnips, the most voluminous of all foods, are used along with oil-cake and bean-meal, and if from any circumstances it becomes necessary to replace a large amount of the former by either of the latter substances, the deficient bulk must be replaced by hay or straw.

It has been already remarked that there are three great purposes to which the food consumed is appropriated; the increase of weight of the animal--the object the feeder has in view and desires to promote--the supplying the waste of the tissues, and the process of respiration, both of which are sources of loss of food, and which it must necessarily be his aim to diminish as much as possible. The circumstances which must be attended to in order to do this are sufficiently well understood. It has been clearly established that the natural heat of the animal is sustained by the consumption of a certain quantity of its food in the respiratory process, during which it undergoes exactly the same changes as those which occur during combustion. It has further been observed, that the temperature of the body remains unchanged, whatever be that of the surrounding air; and it is obvious that if it is to continue the same in winter as in summer, a larger quantity of fuel (i. e. food) must be consumed for this purpose, just as a room requires more fire to keep it warm in winter than in summer, and hence it naturally follows, that if the animal be kept in a warm locality the food is economized. It may also be inferred that, if it were possible, consistently with the health of the animal, to keep it in a room artificially heated to the temperature of its own body, this source of waste of food would be entirely removed. It is not possible, however, to do this, because a limit is set to it by physiological laws, which cannot be infringed with impunity; but the housing of cattle, so as to diminish this waste as far as possible, is a point in regard to the propriety of which theory and practice are at one.

The old feeders kept their cattle in large open courts, where they were exposed to every vicissitude of the weather, but as intelligence advanced, we find them substituting, first hammels, and then stalls, in which the animals are kept during the whole time of fattening at an equable temperature. The effect of this is necessarily to introduce a considerable economy of the food required to sustain the animal heat; but it also produces a saving in another way, for it diminishes the waste of the tissues.

It has been ascertained by accurate experiments made chiefly on man, that muscular exertion is one of the most important causes of the waste of the tissues, and of increased respiratory activity. We cannot move a limb without producing a corresponding consumption of matters already laid up within the body; and it has also been found, that the difference in the quantity of carbonic acid expired during rest and active exertion, is very large. The inference to be drawn from this is, that when it is sought to fatten an animal rapidly, every effort must be made to restrain muscular motion so far as compatible with health. Hence, the peculiar advantage of stall-feeding, in which the animal is confined to one spot, and the more thoroughly it can be kept still, the greater will be the economy of food. This is gained by darkening the house, and excluding all persons, except when their presence is indispensable.

An extension of the same principle has led to the use of food artificially heated, but it is doubtful whether the advantages derived from it are commensurate to the increased expense of the process; at least opinions differ among the best informed practical men on this subject.

Many other matters, besides these mentioned, exercise an important influence on the feeding of stock, such as the general health of the animal, the breed, etc. These are subjects, however, which bear more directly on practical agriculture, and need not be discussed here.

The judicious feeder will not only give due weight to the principles already discussed in all he does, but he must take into consideration the extent to which they are liable to be modified in particular cases. He must also attend to the cost of different kinds of food, and the value of the manure produced by them, subjects of much importance in a practical point of view, and which must influence him greatly in choice of the particular substances he supplies to his cattle.

###

www.ingramcontent.com/pod-product-compliance
Lightning Source LLC
Chambersburg PA
CBHW07085418052 6
45168CB00005B/1811